HANDBOOK OF ELECTRONIC MATERIALS
Volume 1

HANDBOOK OF ELECTRONIC MATERIALS

Compiled by:
ELECTRONIC PROPERTIES INFORMATION CENTER
Hughes Aircraft Company
Culver City, California

Sponsored by:
AIR FORCE MATERIALS LABORATORY
Air Force Systems Command
Wright Patterson Air Force Base, Ohio

Volume 1:
OPTICAL MATERIALS PROPERTIES, 1971

Volume 2:
III-V SEMICONDUCTING COMPOUNDS, 1971

Volume 3:
SILICON NITRIDE FOR MICROELECTRONIC APPLICATIONS, PART I:
                                              PREPARATION AND PROPERTIES, 1971

In preparation:

Volume 4:
NIOBIUM ALLOYS AND COMPOUNDS

Volume 5:
GROUP IV SEMICONDUCTING COMPOUNDS

Volume 6:
SILICON NITRIDE FOR MICROELECTRONIC APPLICATIONS, PART II: APPLICATIONS

HANDBOOK OF ELECTRONIC MATERIALS
Volume 1

Optical Materials Properties

A. J. Moses
Electronic Properties Information Center
Hughes Aircraft Company, Culver City, California

IFI/PLENUM · NEW YORK-WASHINGTON-LONDON · 1971

This document has been approved for public release and sale;
its distribution is unlimited. Sponsored by: Air Force Materials
Laboratory, Wright-Patterson Air Force Base, Ohio.

Library of Congress Catalog Card Number 76-147312
SBN 306-67101-8

©1971   IFI/Plenum Data Corporation, a Subsidiary of
Plenum Publishing Corporation
227 West 17th Street, New York, N.Y. 10011
All Rights Reserved

No part of this publication may be reproduced in any form
without written permission from the Publisher.

Printed in the United States of America

FOREWORD

This report was prepared by Hughes Aircraft Company, Culver City, California under Contract Number F33615-70-C-1348. The work was administered under the direction of the Air Force Materials Laboratory, Air Force Systems Command, Wright-Patterson Air Force Base, Ohio, with Mr. B. Emrich, Project Engineer.

The Electronic Properties Information Center (EPIC) is a designated Information Analysis Center of the Department of Defense authorized to provide information to the entire DOD community. The purpose of the Center is to provide a highly competent source of information and data on the electronic, optical and magnetic properties of materials of value to the Department of Defense. Its major function is to evaluate, compile and publish the experimental data from the world's unclassified literature concerned with the properties of materials. All materials relevant to the field of electronics are within the scope of EPIC: insulators, semiconductors, metals, superconductors, ferrites, ferroelectric, ferromagnetics, electroluminescents, thermionic emitters and optical materials. The Center's scope includes information on over 100 basic properties of materials; information generally regarded as being in the area of devices and/or circuitry is excluded.

CONTENTS

INTRODUCTION - Preparation of Data Sheets 1

DATA SHEETS

    Alumina (Sapphire) 4
    Aluminum (Film) 6
    Ammonium Dihydrogen Phosphate (ADP) 8
    Arsenic Trisulfide (Glass) 10
    Barium Fluoride 12
    Beryllium Oxide 14
    Cadmium Sulfide 16
    Cadmium Telluride 18
    Calcium Carbonate (Calcite) 20
    Calcium Fluoride 22
    Cesium Bromide 24
    Cesium Iodide 26
    Copper (Film) 28
    Cuprous Chloride 30
    Gallium Arsenide 32
    Germanium 34
    Gold (Film) 36
    Indium Arsenide 38
    Lead Sulfide 40
    Lithium Fluoride 42
    Lithium Niobate 44
    Magnesium Fluoride (Film) 46
    Magnesium Fluoride (Single Crystal) 48
    Magnesium Oxide 50
    Palladium (Film) 52
    Platinum (Film) 54
    Potassium Bromide 56
    Potassium Chloride 58
    Potassium Iodide 60
    Selenium (Amorphous Film) 62
    Selenium (Hexagonal) 64
    Silica (Crystalline) 66
    Silica (Fused) 68
    Silicon 70
    Silicon Carbide (Type 6H) 72
    Silver (Film) 74
    Silver Bromide 76
    Silver Chloride 78
    Sodium Chloride 80
    Sodium Fluoride 82
    Strontium Titanate 84
    Tellurium (Polycrystalline Film) 86
    Tellurium (Single Crystal) 88
    Thallium Bromoiodide (KRS-5) 90
    Thallium Chlorobromide (KRS-6) 92

DATA SHEETS (Continued)

    Titanium 94
    Titanium Dioxide (Rutile) 96
    Zinc Selenide (Cubic) 98
    Zinc Sulfide (Cubic) 100

APPENDIX A

    Wavelength Conversion Factors 102

APPENDIX B

    Glossary of Optical Terms 104

INTRODUCTION

The designer of optical systems must base the selection of optical materials on a knowledge of optical, physical, thermal and mechanical properties. Frequently, the selection of the materials must result from a trade-off analysis as no one material has an ideal set of properties. This compilation of Optical Materials Properties Data Sheets summarizes the vital properties, needed to select an optical material for use in the ultraviolet, visible and infrared regions of the electromagnetic spectrum. Forty-nine materials are described in this compilation and each sheet of two pages presents the properties of a particular material. The listed properties are:

Physical Properties:	Density
	Melting/Softening Temperature
	Solubility in Water at Room Temperature
Thermal Properties:	Linear Expansion Coefficient
	Thermal Conductivity
	Specific Heat
Mechanical Properties:	Young's Modulus
	Hardness
Optical Properties:	Transmittance
	Reflectance
	Refractive Index
	Extinction Coefficient or Absorption Coefficient
	Dispersion Coefficient
	Dispersion Equation

In some cases, even a thorough search of the literature did not yield data considered suitable for the data sheet and other useful plots are offered in place thereof. Most optical data are presented in graphical form as a function of wavelength. In some cases, wavenumber or photon energy are the independent variables, but conversion tables (Appendix A) will facilitate the use of these graphs. Optical terms are defined in Appendix B.

The computerized data retrieval system of the Electronic Properties Information Center, EPIC, provided most of the references for the data sheets (Figure 1). The EPIC system contains over 42,000 documents and covers 100 properties of more than 7,000 materials. The literature includes: (1) technical journals, (2) technical reports, (3) textbooks, (4) indexing and abstracting services, (5) theses and dissertations, (6) patents, (7) conference and symposium proceedings, (8) standards and specification documents, (9) vendor and trade literature, (10) bibliographies, and (11) private communications. This literature is in the following languages:

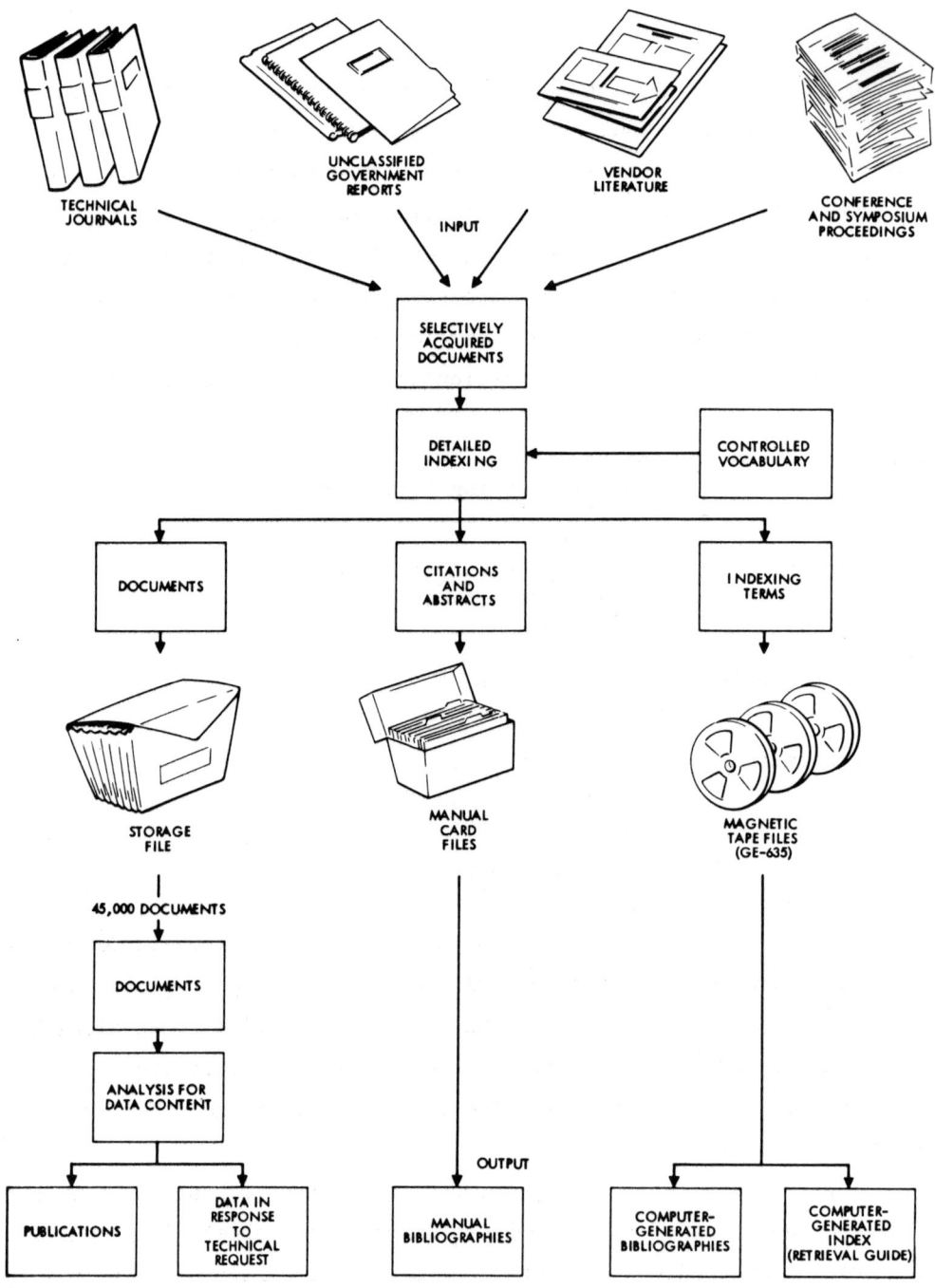

FIGURE 1

EPIC INFORMATION PROCESSING SYSTEM

English, German, French and Russian. The growth of the optical literature is illustrated by an over 200% increase in papers containing refractive index measurements over the seven year span from 1961 to 1968. The Electronic Properties Information Center, sponsored by the Air Force Materials Laboratory, includes optical materials properties in its coverage and its <u>technical information analysis specialists</u> can provide valuable aid to designers, scientists and engineers by assimilating and packaging pertinent information from this vast body of literature in the form of <u>books</u> and <u>reports</u> and the EPIC <u>Technical Answering Service</u>.

Technical Answering Service responds to inquiries ranging in complexity from simple requests for data point values to requests for comprehensive reviews of the literature. This service is available to U.S. Government agencies, their contractors, subcontractors, suppliers, and those in a position to support the defense effort. Inquiries may be directed to:

> Electronic Properties Information Center
> Hughes Aircraft Company
> Bldg. 6: E-148
> Centinela and Teale Streets
> Culver City, California 90230
> Telephone: (213) 391-0711, Ext. 6596

The EPIC Bulletin, published quarterly, announces new publications and current activities of the Center. Users may request receipt of the Bulletin on a regular basis.

ALUMINA

OPTICAL MATERIALS PROPERTIES MATERIAL: ALUMINA
DATA SHEET   (Sapphire)

INTRODUCTION: This data sheet provides information on single crystal synthetic sapphire. The optical data are for the ordinary ray.

PHYSICAL PROPERTIES, (298°K)

Density, (g/cm³) __3.98__

Melting/Softening Temp. (°K) __2303__

Solubility in Water, (g./100 g. H_2O) __9.8×10^{-5}__

MECHANICAL PROPERTIES*, (298°K)

Young's Modulus, (psi) __50×10^6__

Hardness, (Knoop) __2000, (1000 g. indenter)__

THERMAL PROPERTIES*, (298°K)

Linear Expansion Coeff., (°K)⁻¹ __5.8×10^{-6}__

Thermal Conductivity (10^{-4} cal/(cm sec °K) __0.10__

Specific Heat, (cal/g)/°K __0.18__

*60° orientation to optic axis

OPTICAL PROPERTIES, (298°K)

Dispersion Equation

$$n^2 - 1 = \sum_i \frac{A_i \lambda^2}{\lambda^2 - \lambda_t^2}$$

λ_1 = 0.06144821 λ_1^2 = 0.00377588 A_1 = 1.023798
λ_2 = 0.1106997 λ_2^2 = 0.0122544 A_2 = 1.058264
λ_3 = 17.92656 λ_3^2 = 321.3616 A_3 = 5.280792

(0.26 - 5.6 μ; λ in μ.)

Transmission Region, (External Transmittance ≥ 10% with __2.0__ mm. thickness) __0.15 - 6.5 μ__

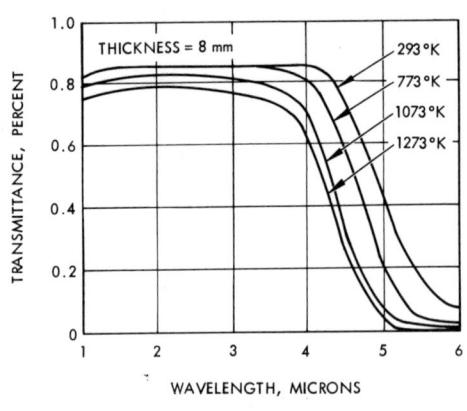

(REF. 1)

MATERIAL: ALUMINA
(Sapphire)

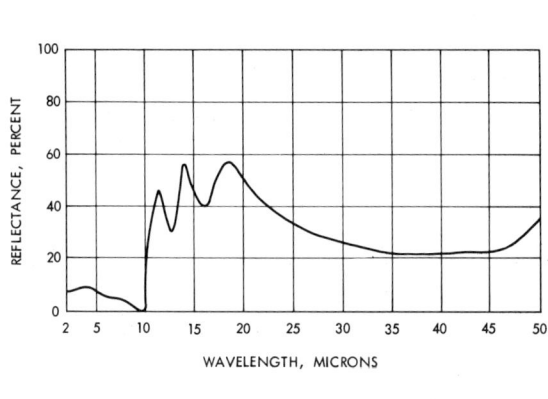
(REF. 2)

(REF. 3)

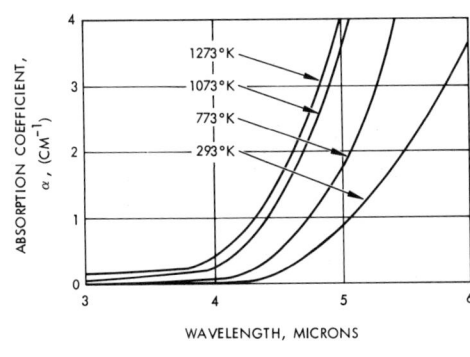
(REF. 1)

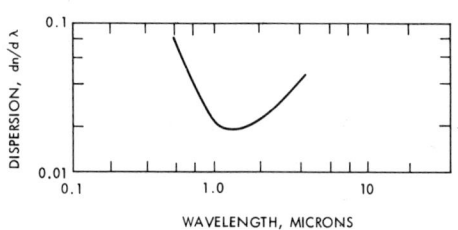
(REF. 4)

REFERENCES:

1. U. P. Oppenheim, and U. Even, J. Opt. Soc. Am., 52, 1078-9, (1962).

2. D. E. McCarthy, Appl. Optics, 2, 591-595, (1963).

3. I. H. Malitson, et al, J. Opt. Soc. Am., 48, 72-73, (1958).

4. W. L. Wolfe, Ed., "Handbook of Military Infrared Technology," U.S. Govt. Printing Office, Washington, (1965).

ALUMINUM

OPTICAL MATERIALS PROPERTIES DATA SHEET   MATERIAL: <u>ALUMINUM</u>
<u>(Film)</u>

INTRODUCTION: This data sheet presents data for aluminum film, formed by evaporation. The optical data are compromised by uncertainty concerning the crystallinity of the film.

PHYSICAL PROPERTIES, (298°K)	
Density, (g/cm^3)	2.70
Melting/Softening Temp. (°K)	933
Solubility in Water, (g./100 g. H$_2$O)	0.000

MECHANICAL PROPERTIES, (298°K)	
Young's Modulus, (psi)	7.06 x 10^5
Hardness	~25 Brinnell

THERMAL PROPERTIES, (298°K)	
Linear Expansion Coeff. (°K)$^{-1}$	23.5 x 10^{-6}
Thermal Conductivity (10^{-4} cal/(cm sec °K)	0.504
Specific Heat, (cal/g)/°K	0.214

OPTICAL PROPERTIES, (298°K)	
Dispersion Equation:	not available
Transmission Region, (External Transmittance ≥10% with _____ mm. thickness)	not available

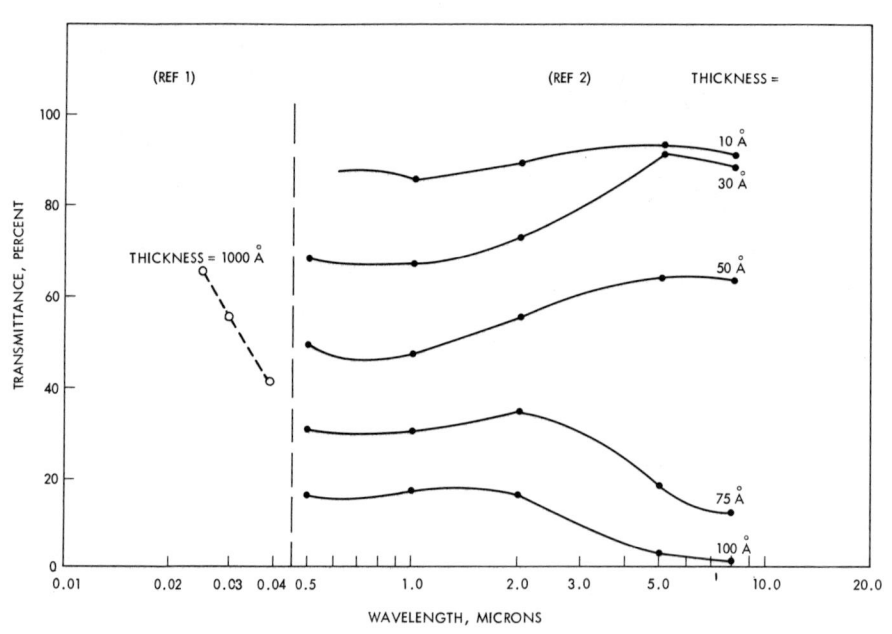

MATERIAL: ALUMINUM
(Film)

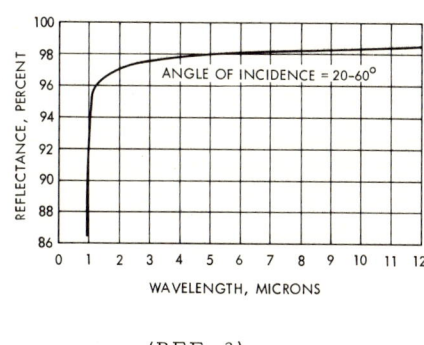

(REF. 3)

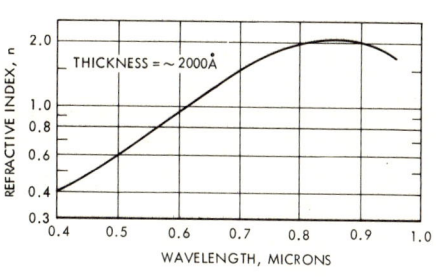

(REF. 4)

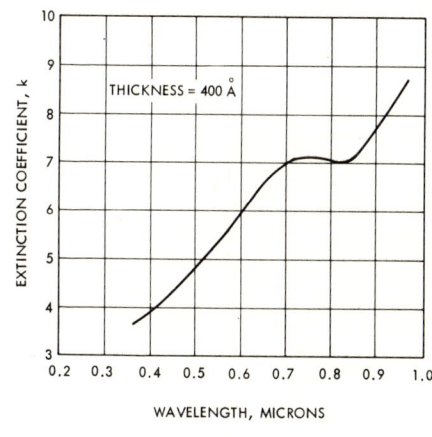

(REF. 5)

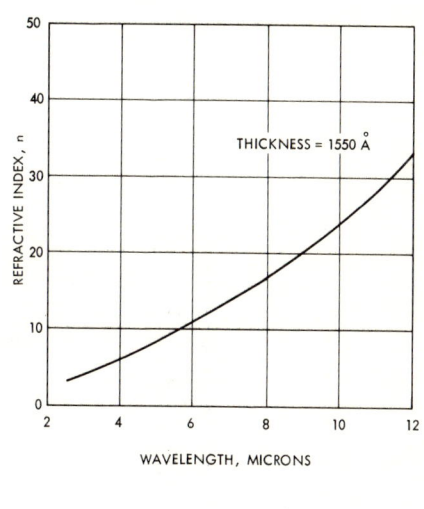

(REF. 6)

REFERENCES:

1. O. P. Rustgi, J. Opt. Soc. Am., 55, 630-634, (1965).
2. H. Mayer, "Physik duenner Schichten", Pt. I., Wiss. Verlagsgesell., Stuttgart, (1950).
3. D. M. Gates, et al., J. Opt. Soc. Am., 48, 88-89, (1958).
4. L. G. Schulz, and F. R. Tangherlini, J. Opt. Soc. Am., 44, 362-368, (1954).
5. L. G. Schulz, J. Opt. Soc. Am., 44, 357-362, (1954).
6. J. R. Beattie, Phil. Mag., 46, 235-245, (1955).

AMMONIUM DIHYDROGEN PHOSPHATE

OPTICAL MATERIALS PROPERTIES DATA SHEET   MATERIAL: AMMONIUM DIHYDROGEN PHOSPHATE (ADP)

INTRODUCTION: This data sheet contains information on single crystal ammonium dihydrogen phosphate.

PHYSICAL PROPERTIES, (298°K)
- Density, (g/cm^3)   1.803
- Melting/Softening Temp. (°K)   526
- Solubility in Water, (g./100 g. H$_2$O)   22.7 (273°K)

MECHANICAL PROPERTIES, (298°K)
- Young's Modulus, (psi)   Not available
- Hardness, (Knoop)   Not available

THERMAL PROPERTIES, (298°K)
- Linear Expansion Coeff., (°K)$^{-1}$ 39×10^{-6} (a axis); 1.9×10^{-6} (c axis)
- Thermal Conductivity (10^{-4} cal/(cm sec °K))   17 (c axis); 30 (a axis)
- Specific Heat, (cal/g)/°K   0.29

OPTICAL PROPERTIES, (298°K)
- Dispersion Equation   Not available
- Transmission Region, (External Transmittance ≥10% with 2.0 mm. thickness)   0.13 - 1.7µ

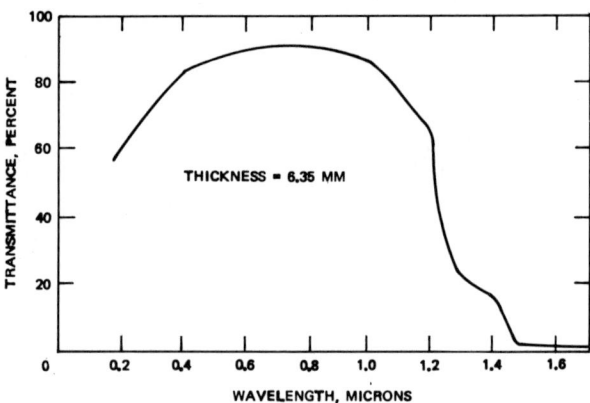

(REF. 1)

MATERIAL: AMMONIUM DIHYDROGEN PHOSPHATE (ADP)

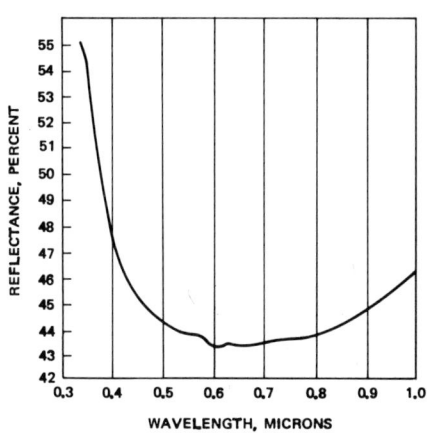

(REF. 2)

(REF. 3)

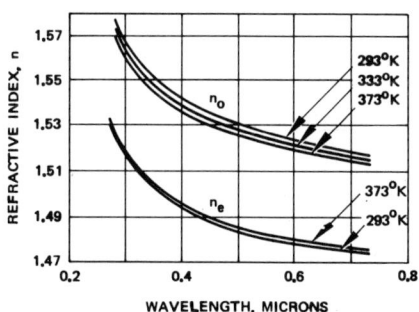

(REF. 4)

REFERENCES:

1. G.D. Burnett, Electronic Industries, 21, 90-95, (1962).

2. F. Vratny and J.J. Kokalas, Appl. Spectroscopy, 16, 176-184, (1962).

3. F. Zernike, Jr., J. Opt. Soc. Am., 54, 1215-1220, (1964).

4. V.N. Vishneskii, et al, Opt. Spec., (English Transl.), 18, 468-469, (1965).

ARSENIC TRISULFIDE

OPTICAL MATERIALS PROPERTIES DATA SHEET   MATERIAL: __ARSENIC TRISULFIDE (Glass)__

INTRODUCTION: This data sheet presents data for arsenic trisulfide glass.

PHYSICAL PROPERTIES, (298°K)
Density, (g/cm³)   3.198
Melting/Softening Temp. (°K)   483
Solubility in Water, (g./100 g. H_2O)   ~5 x 10⁻⁵ (291°K)

MECHANICAL PROPERTIES, (298°K)
Young's Modulus, (psi) 2.3×10^6
Hardness, (Knoop)   109 (100 g.)

THERMAL PROPERTIES, (298°K)
Linear Expansion Coeff. (°K)⁻¹ 24.6×10^{-6}
Thermal Conductivity (10 cal/(cm sec °K))   4.0 (313°K)
Specific Heat, (cal/g)/°K   not available

OPTICAL PROPERTIES, (298°K)
Dispersion Equation:

$$n^2 - 1 = \sum_{i=1}^{5} \frac{K_i \lambda^2}{\lambda^2 - \lambda_i^2}$$

i	λ_i^2	K_i
1	0.0225	1.8983678
2	0.0625	1.9222979
3	0.1225	0.8765134
4	0.2025	0.1188704
5	750.	0.9569903

Transmission Region, (External Transmittance 10% with __2.0__ mm. thickness)   0.6 - 13µ

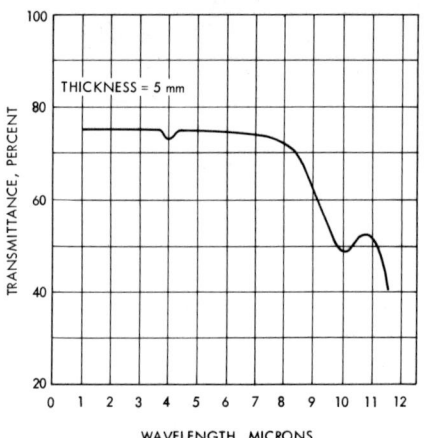

(REF. 1)

MATERIAL: ARSENIC TRISULFIDE (Glass)

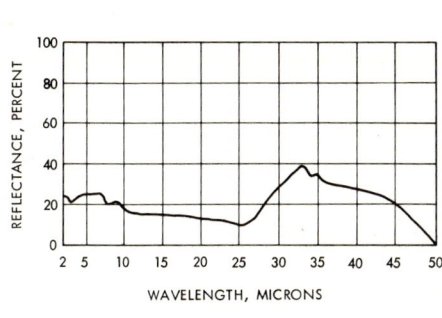

(REF. 2)

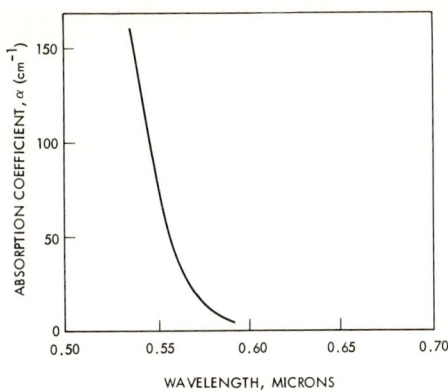

(REF. 4)

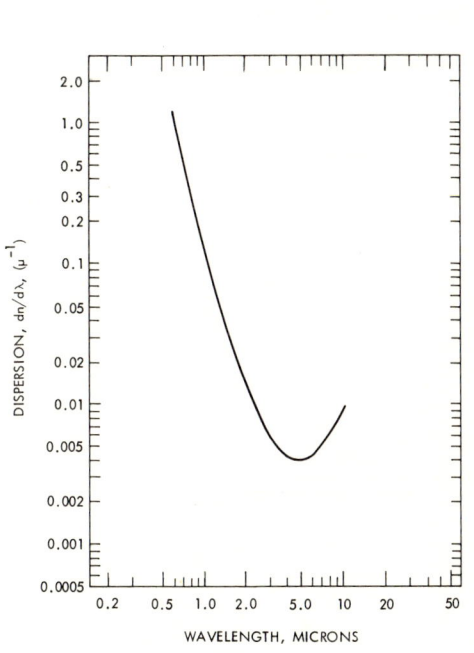

(REF. 3)

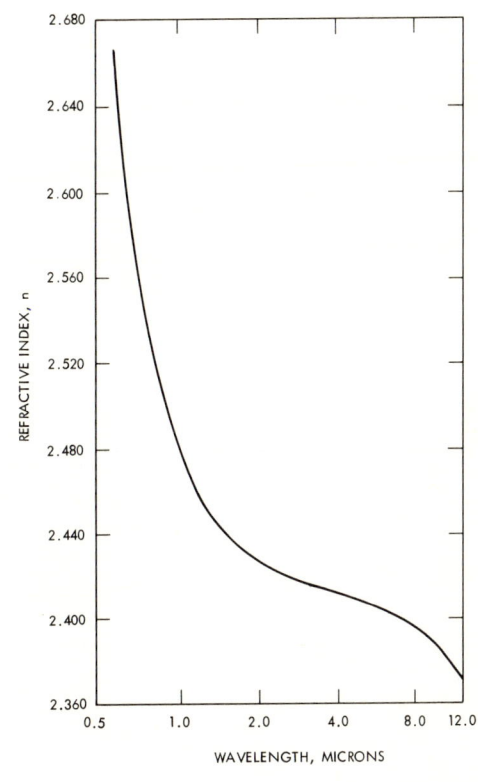

(REF. 3)

REFERENCES:

1. S. S. Ballard, et al., "Optical Materials for Infrared Instrumentation," AD 217 367, (1959).

2. D. E. McCarthy, Appl. Optics, 2, 591-595, (1963).

3. W. S. Rodney, et al., J. Opt. Soc. Am., 48, 633-636, (1958).

4. G. Getov, et al., Phys. Stat. Sol., 21, K87-K89, (1967).

BARIUM FLUORIDE

OPTICAL MATERIALS PROPERTIES
DATA SHEET

MATERIAL: BARIUM FLUORIDE

INTRODUCTION: This data sheet provides information for synthetic single crystal barium fluoride.

PHYSICAL PROPERTIES, (298°K)
- Density, (g/cm^3)   4.83
- Melting/Softening Temp. (°K)   1553
- Solubility in Water, (g./100 g. H$_2$O)   0.16

MECHANICAL PROPERTIES, (298°K)
- Young's Modulus, (psi) 7.7×10^{-6}
- Hardness, (Knoop)   82, (500 g.)

THERMAL PROPERTIES, (298°K)
- Linear Expansion Coeff. (°K)$^{-1}$ 18.4×10^{-6}
- Thermal Conducitivity (10^{-4} cal/(cm sec °K))   280
- Specific Heat, (cal/g)/°K   not available

OPTICAL PROPERTIES, (298°K)

Dispersion Equation:

$$n^2 - 1 = \frac{0.643356\lambda^2}{\lambda^2 - (0.057789)^2} + \frac{0.50676\lambda^2}{\lambda^2 - (0.10968)^2} + \frac{3.8261\lambda^2}{\lambda^2 - (46.3864)^2}$$

Transmission Region, (External Transmittance ≥ 10% with 2.0 mm. thickness)   0.25 - 15 µ

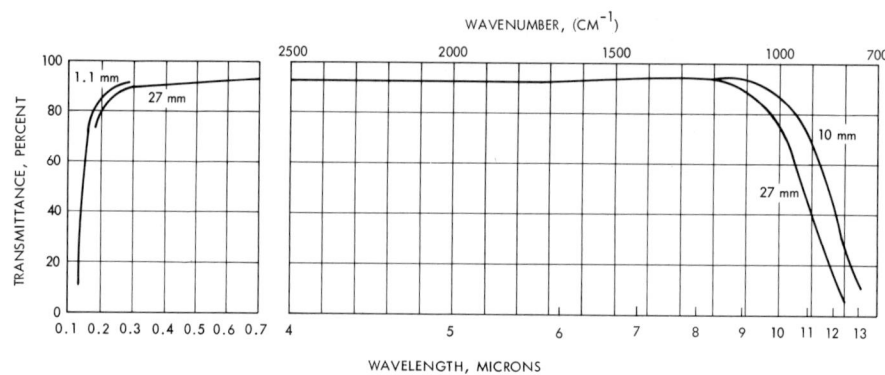

(REF. 1)

MATERIAL: **BARIUM FLUORIDE**

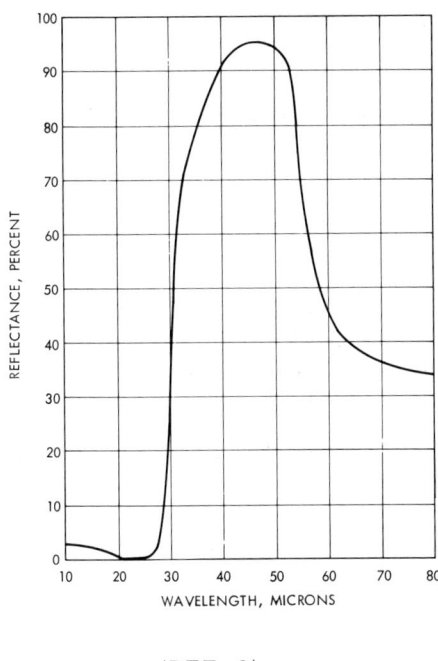

(REF. 2)

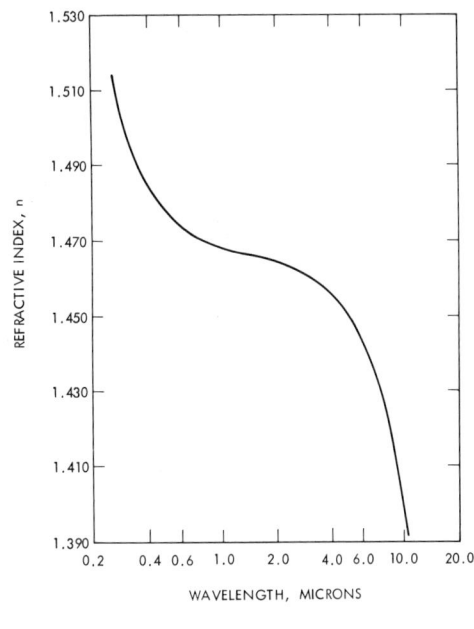

(REF. 3)

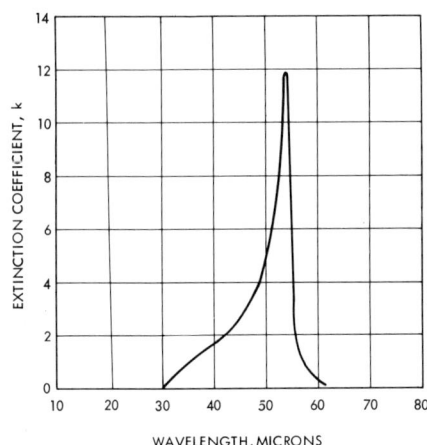

(REF. 2)

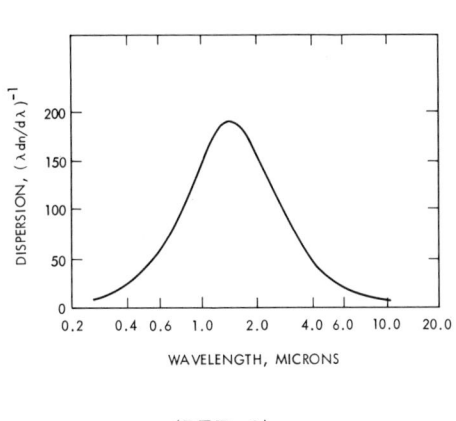

(REF. 3)

REFERENCES:

1. A. Smakula, et al., "Harshaw Optical Crystals", Harshaw Chemical Co., Cleveland, (1967).

2. W. Kaiser, et al., Phys. Rev., 127, 1950-1954, (1962).

3. I. H. Malitson, J. Opt. Soc. Am., 54, 628 - 632, (1964).

BERYLLIUM OXIDE

OPTICAL MATERIALS PROPERTIES DATA SHEET   MATERIAL: BERYLLIUM OXIDE

INTRODUCTION: This data sheet contains information for single crystal beryllium oxide.

PHYSICAL PROPERTIES, (298°K)
- Density, (g/cm^3)   3.00
- Melting/Softening Temp. (°K)   2823
- Solubility in Water, (g./100 g. H$_2$O)   <0.01

MECHANICAL PROPERTIES, (298°K)
- Young's Modulus, (psi) 53×10^6
- Hardness, (Knoop)   917 - 1300

THERMAL PROPERTIES, (298°K)
- Linear Expansion Coeff., (°K)$^{-1}$ 5.5×10^{-6}
- Thermal Conductivity (10^{-4} cal/(cm sec °K))   520
- Specific Heat, (cal/g)/°K   0.24

OPTICAL PROPERTIES, (298°K)
- Dispersion Equation   Not available
- Transmission Region, (External Transmittance ≥10% with _____ mm. thickness)   Not available

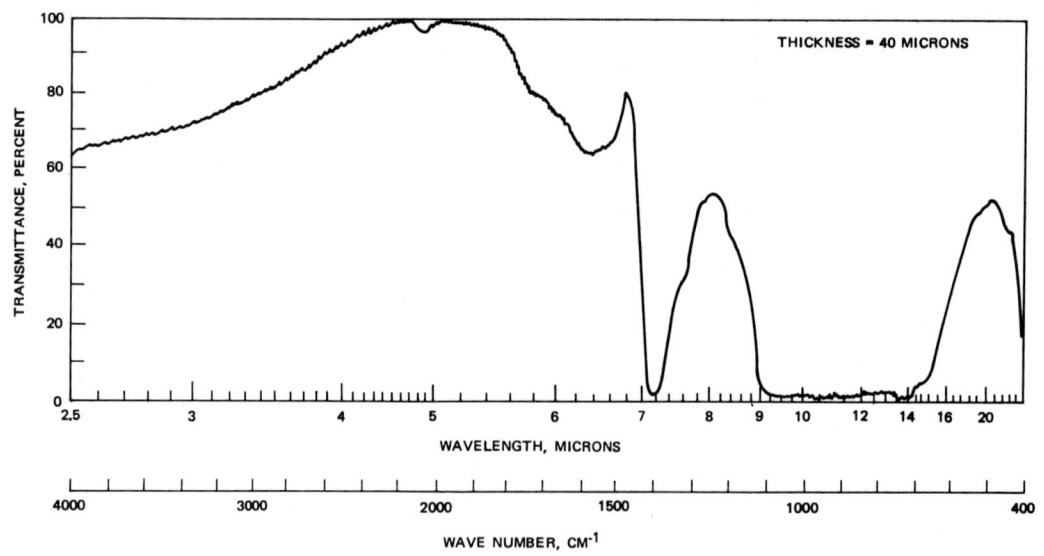

(REF. 1)

MATERIAL: <u>BERYLLIUM OXIDE</u>

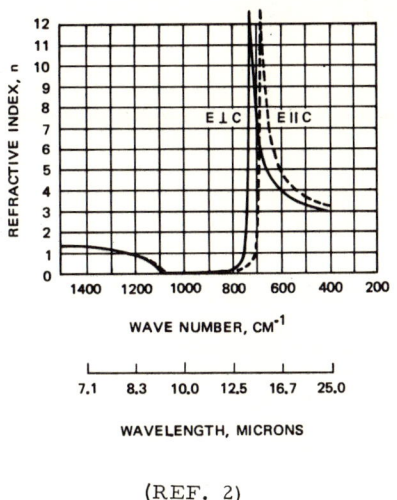

(REF. 2)

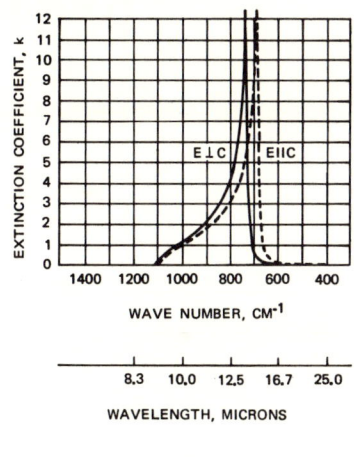

(REF. 2)

(REF. 2)

REFERENCES:

1. R.J. Morrow and H.W. Newkirk, Rev. Int. Hautes Temp. Refractaires, <u>6</u>, No. 2, 99-104, (1969).
2. E. Loh, Phys. Rev., <u>166</u>, 673-678, (1968).

CADMIUM SULFIDE

OPTICAL MATERIALS PROPERTIES
DATA SHEET

MATERIAL: CADMIUM SULFIDE

INTRODUCTION: This data sheet summarizes the properties of bulk, hexagonal single crystal cadmium sulfide.

PHYSICAL PROPERTIES, (298°K)

Density, (g/cm³)   4.82

Melting/Softening Temp. (°K)   1253 (subl. at 1 atm.)

Solubility in Water, (g./100 g. H₂O) 1.3×10^{-4} (291°K)

MECHANICAL PROPERTIES, (298°K)

Young's Modulus, (psi)   not available

Hardness, (Knoop)   121, (100 g) (∥ to C axis)

THERMAL PROPERTIES, (298°K)

Linear Expansion Coeff. (°K)⁻¹ 5×10^{-6} (⊥ C axis) 3.5×10^{-6} (∥ C axis)

Thermal Conductivity (10^{-4} cal/(cm sec °K))   380

Specific Heat, (cal/g)/°K   0.088 (273°K)

OPTICAL PROPERTIES, (298°K)

Dispersion Equation

$$n_o^2 = 5.235 + \frac{1.819 \times 10^7}{\lambda^2 - 1.651 \times 10^7}$$

$$n_e^2 = 5.239 + \frac{2.076 \times 10^7}{\lambda^2 - 1.651 \times 10^7}$$

Transmission Region, (External Transmittance ≥ 10% with 2.0 mm. thickness)   0.5 - 16μ

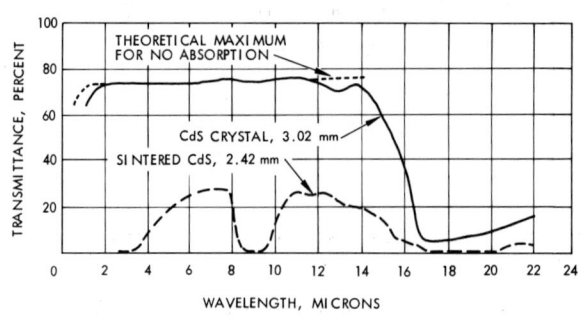

(REF. 1)

MATERIAL: CADMIUM SULFIDE

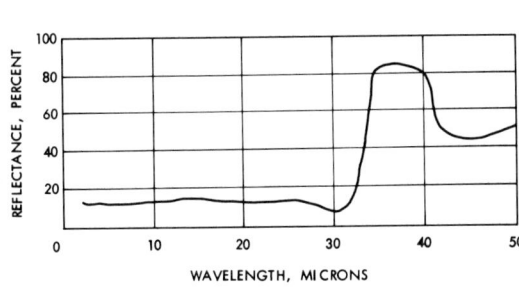

(REF. 2)

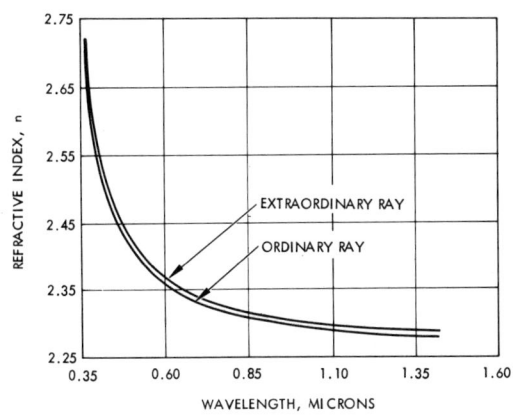

(REF. 3)

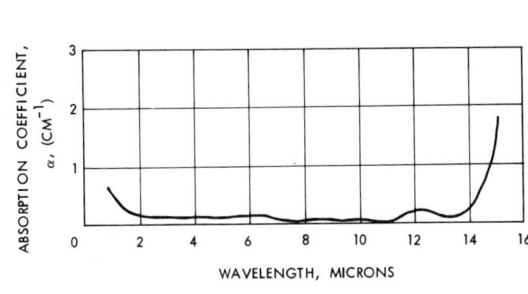

(REF. 1)

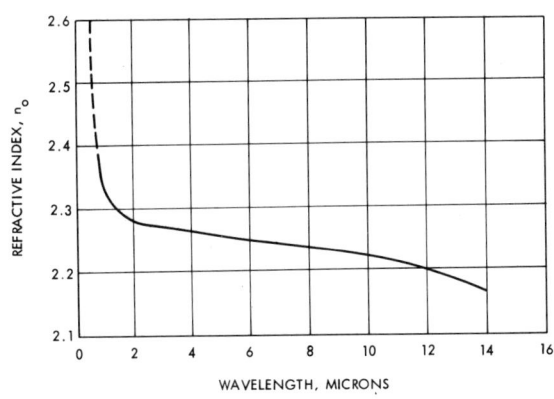

(REF. 1)

REFERENCES:

1. A. B. Francis, and A. I. Carlson, J. Opt. Soc. Am., 50, 118-121, (1960).
2. D. E. McCarthy, Applied Optics, 7, 1997-2000, (1968).
3. T. M. Bieniewski, and S. J. Czyzak, J. Opt. Soc. Am., 53, 649-497, (1963).

CADMIUM TELLURIDE

OPTICAL MATERIALS PROPERTIES   MATERIAL: <u>CADMIUM TELLURIDE</u>

INTRODUCTION: <u>This data sheet presents data for hot-pressed polycrystalline cadmium telluride.</u>

PHYSICAL PROPERTIES, (298°K)
- Density, (g/cm^3) _____ 5.85 _____
- Melting/Softening Temp. (°K) _____ 1318 _____
- Solubility in Water, (g./100 g. H$_2$O) _____ not available _____

MECHANICAL PROPERTIES, (298°K)
- Young's Modulus, (psi) _____ 5.3×10^6 _____
- Hardness, (Knoop) _____ 45 _____

THERMAL PROPERTIES, (298°K)
- Linear Expansion Coeff. (°K) _____ 5.5×10^{-6} _____
- Thermal Conductivity (10^{-4} cal/(cm sec °K)) _____ 98 _____
- Specific Heat, (cal/g)/°K _____ not available _____

OPTICAL PROPERTIES, (298°K)
- Dispersion Equation: _____ not available _____
- Transmission Region, (External Transmittance ≥10% with _____ 2.0 _____ mm. thickness) _____ 0.9 - 16μ _____

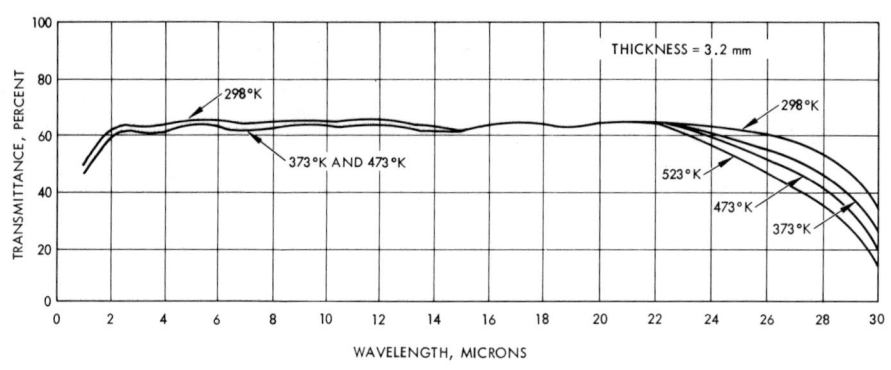

(REF. 1)

MATERIAL: CADMIUM TELLURIDE

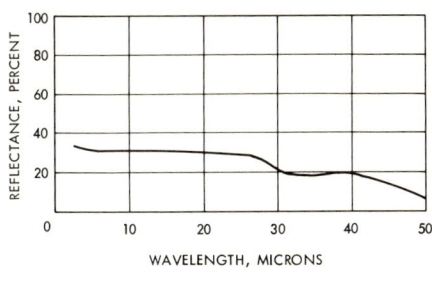

(REF. 2)

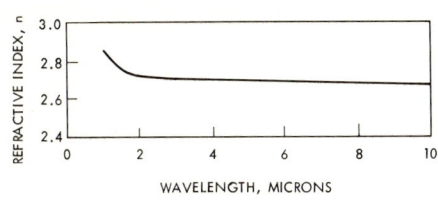

(REF. 3)

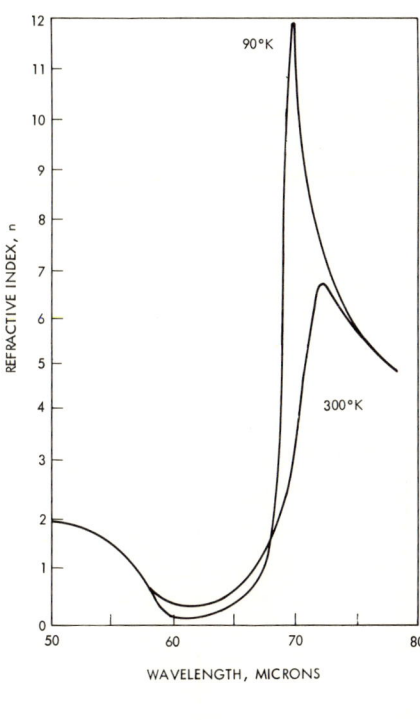

(REF. 4)

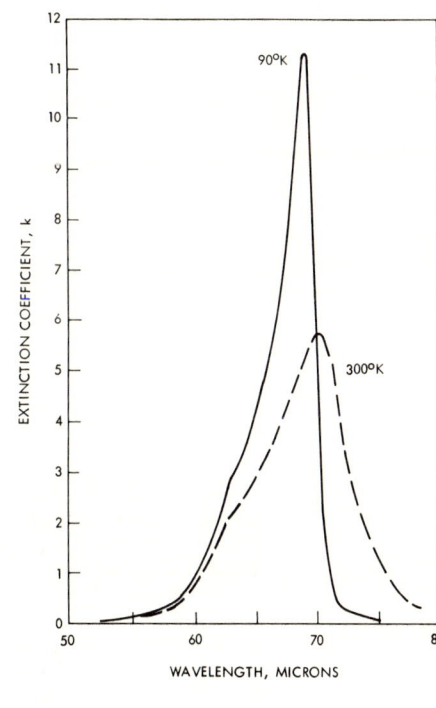

(REF. 4)

REFERENCES:

1. L. S. Ladd, Infrared Physics, 6, 145-151, (1966).
2. D. E. McCarthy, Appl. Optics, 7, 1997-2000, (1968).
3. Eastman Kodak Company, Kodak IRTRAN-6 Material Data Sheet.
4. A. Mitsuishi, J. Phys. Soc. Japan, 16, 533-537, (1961).

CALCIUM CARBONATE

OPTICAL MATERIALS PROPERTIES
DATA SHEET

MATERIAL: CALCIUM CARBONATE (Calcite)

INTRODUCTION: This data sheet provides data for single crystal calcium carbonate.

PHYSICAL PROPERTIES, (298°K)
- Density, (g/cm^3)   2.71
- Melting/Softening Temp. (°K)   1167 (Dissocn)
- Solubility in Water, (g./100 g. H$_2$O)   0.0014

MECHANICAL PROPERTIES*, (298°K)
- Young's Modulus, (psi)   10.5, 12.8
- Hardness, (Mohs)   3

THERMAL PROPERTIES*, (273°K)
- Linear Expansion Coeff., (°K)$^{-1}$ 25, (-) 5.8
- Thermal Conductivity (10^{-4} cal/(cm sec °K))   132, 111
- Specific Heat, (cal/g)/°K   0.203

OPTICAL PROPERTIES, (298°K)
- Dispersion Equation   Not available
- Transmission Region, (External Transmittance ≥10% with 2.0 mm. thickness)   0.2 - 5.5μ

*E ∥ C and E ⊥ C respectively

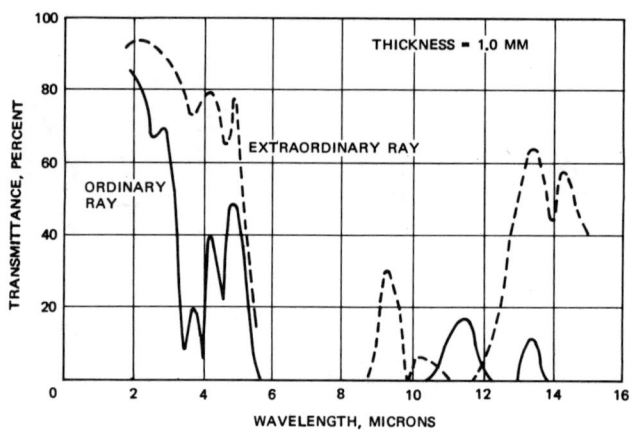

(REF. 1)

MATERIAL: CALCIUM CARBONATE (Calcite)

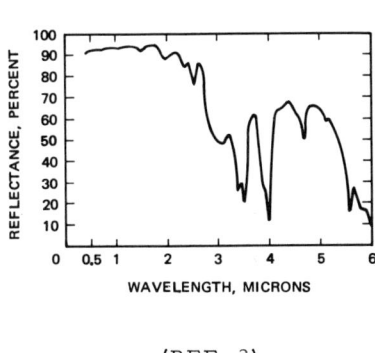

(REF. 2)

(REF. 1)

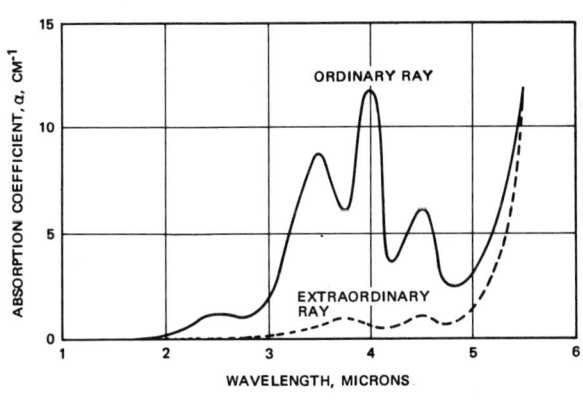

(REF. 3)

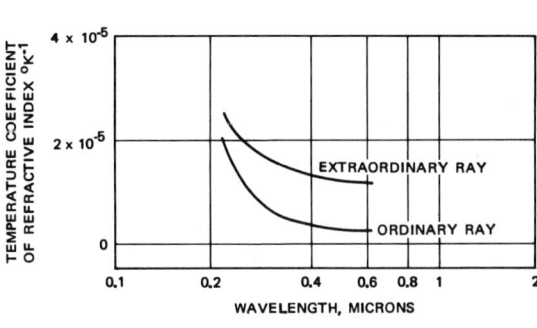

(REF. 3)

REFERENCES:

1. S.S. Ballard, et al., "Optical Materials for Infrared Instrumentation," Report No. AD 217367, (1959).
2. W.A. Hovis, Jr., Appl. Optics, 5, 245-248, (1966).
3. A. Smakula, Opt. Acta, 9, 205-222, (1962).

CALCIUM FLUORIDE

OPTICAL MATERIALS PROPERTIES DATA SHEET

MATERIAL: CALCIUM FLUORIDE

INTRODUCTION: This data sheet presents information for single crystal calcium fluoride.

PHYSICAL PROPERTIES, (298°K)

    Density, (g/cm³)   3.179

    Melting/Softening Temp. (°K)   1613

    Solubility in Water, (g./100 g. H₂O)   0.0017

MECHANICAL PROPERTIES, (298°K)

    Young's Modulus, (psi) 11.0×10^{-6}

    Hardness, (Knoop)   158.3 (500 g)

THERMAL PROPERTIES, (298°K)

    Linear Expansion Coeff. (°K)⁻¹ 24×10^{-6}

    Thermal Conductivity (10^{-4} cal/(cm sec °K))   232 (309°K)

    Specific Heat, (cal/g)/°K   0.204 (273°K)

OPTICAL PROPERTIES, (298°K)

Dispersion Equation:

$$n^2 - 1 = \frac{0.5675888\lambda^2}{\lambda^2 - (0.050263605)^2} + \frac{0.4710914\lambda^2}{\lambda^2 - (0.1003909)^2} + \frac{3.8484723\lambda^2}{\lambda^2 - (34.649040)^2}$$

Transmission Region, (External Transmittance ≥ 10% with 2.0 mm. thickness)   0.13 - 12 μ

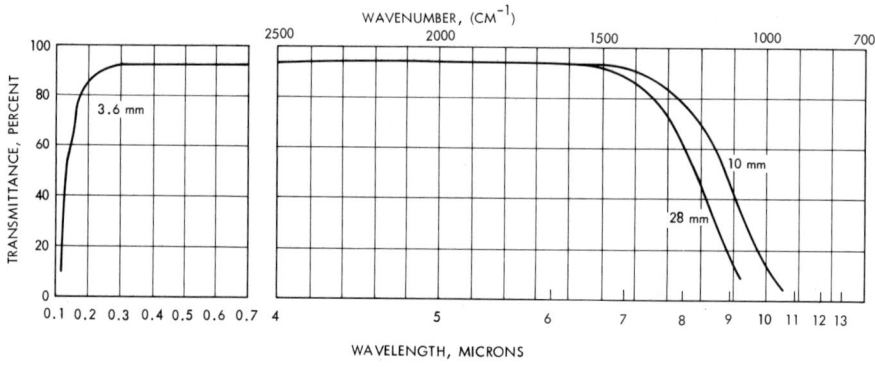

(REF. 1)

MATERIAL: CALCIUM FLUORIDE

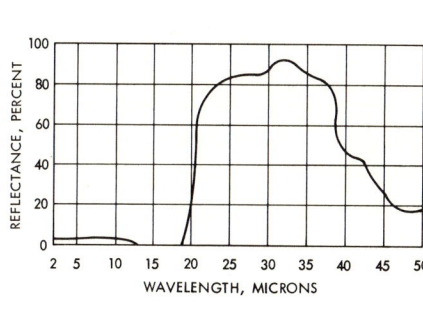

(REF. 2)

(REF. 3)

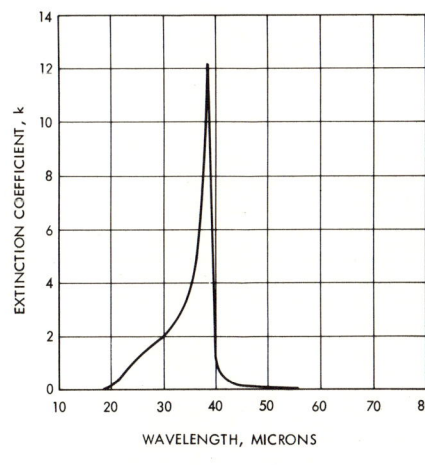

(REF. 4)

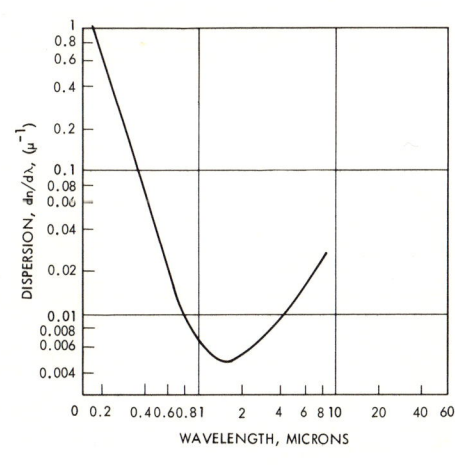

(REF. 5)

REFERENCES:

1. A. Smakula, et al., "Harshaw Optical Crystals," Harshaw Chemical Co., Cleveland, (1967).

2. D. E. McCarthy, Appl. Optics, 2, 591-595, (1963).

3. I. H. Malitson, Appl. Optics, 2, 1103-1107, (1963).

4. W. Kaiser, et al., Phys. Rev., 127, 1950-1954, (1962).

5. A. Smakula, Optica Acta, 9, 205-222, (1962).

CESIUM BROMIDE

OPTICAL MATERIALS PROPERTIES DATA SHEET   MATERIAL: CESIUM BROMIDE

INTRODUCTION: This data sheet presents information on single crystal cesium bromide.

PHYSICAL PROPERTIES, (298°K)
- Density, (g/cm³)   4.44
- Melting/Softening Temp. (°K)   909
- Solubility in Water, (g./100 g. H₂O)   124

MECHANICAL PROPERTIES, (298°K)
- Young's Modulus, (psi) 2.3×10^6
- Hardness, (Knoop)   19.5 (200 g.)

THERMAL PROPERTIES, (298°K)
- Linear Expansion Coeff., (°K)⁻¹ 47.9×10^{-6}
- Thermal Conductivity (10^{-4} cal/(cm sec °K))   0.23
- Specific Heat, (cal/g)/°K   0.063

OPTICAL PROPERTIES, (298°K)

Dispersion Equation:

$$n^2 = 5.640752 - 0.000003338\lambda^2 + \frac{0.0018612}{\lambda^2} + \frac{41110.49}{\lambda^2 - 14390.4} + \frac{0.0290764}{\lambda^2 - 0.024964}$$

Transmission Region, (External Transmittance ≥ 10% with 2.0 mm. thickness)   0.3 - 55μ

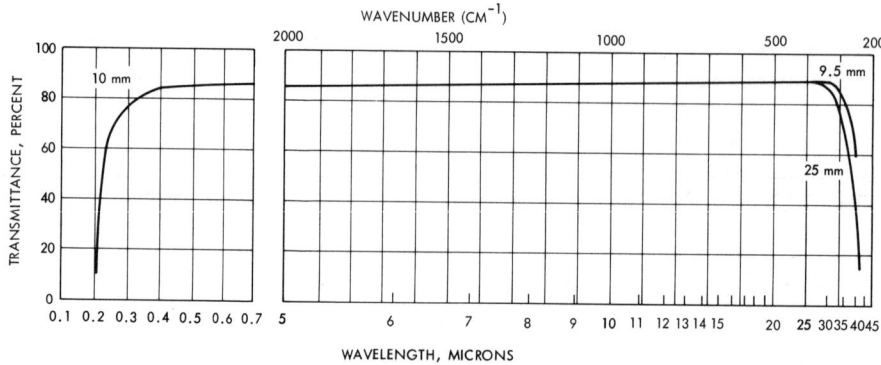

(REF. 1)

MATERIAL: CESIUM BROMIDE

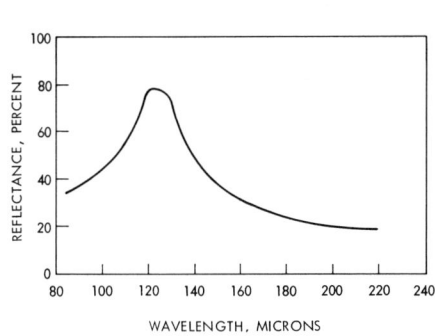

(REF. 2)

(REF. 4)

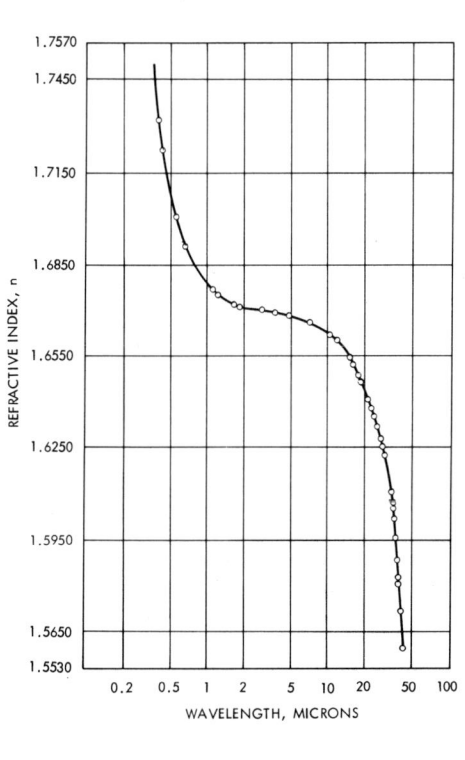

(REF. 3)

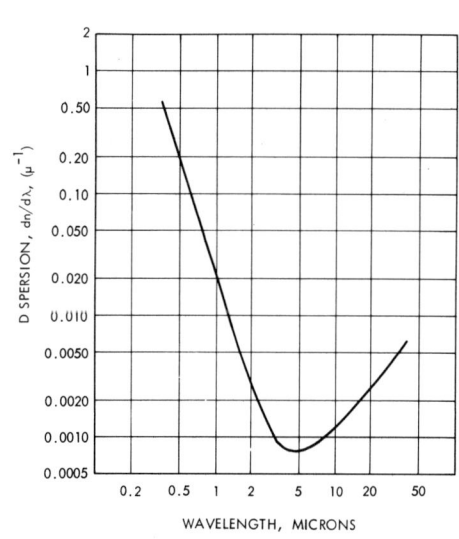

(REF. 3)

REFERENCES:

1. A. Smakula, et al., "Harshaw Optical Crystals," Harshaw Chemical Co., Cleveland, (1967).
2. W. M. Sinton, and W. C. Davis, J. Opt. Soc. Am., 44, 503-504, (1954).
3. W. S. Rodney, and R. J. Spindler, J. S. Res. NBS, 51, 123-126, (1953).
4. E. M. Dianov, Soviet Physics - Solid State, 9, 464-466, (1967).

CESIUM IODIDE

OPTICAL MATERIALS PROPERTIES
DATA SHEET

MATERIAL: CESIUM IODIDE

INTRODUCTION: This data sheet summarizes significant properties for single crystal cesium iodide.

PHYSICAL PROPERTIES, (298°K)

Density, (g/cm³)   4.51

Melting/Softening Temp. (°K)   894

Solubility in Water, (g./100 g. H_2O)   44 (273°K)

MECHANICAL PROPERTIES, (298°K)

Young's Modulus, (psi) 0.769×10^6

Hardness, (Knoop)   not available

THERMAL PROPERTIES, (298°K)

Linear Expansion Coeff. (°K)$^{-1}$ 50×10^{-6}

Thermal Conductivity (10^{-4} cal/(cm sec °K))   27

Specific Heat, (cal/g)/°K   0.048

OPTICAL PROPERTIES, (298°K)

Dispersion Equation:

$$n^2 - 1 = \frac{0.34617251\lambda^2}{\lambda^2 - 0.00052701} + \frac{1.0080886\lambda^2}{\lambda^2 - 0.02149156} + \frac{0.28551800\lambda^2}{\lambda^2 - 0.032761} + \frac{0.39743178\lambda^2}{\lambda^2 - 0.044944} + \frac{3.3605359\lambda^2}{\lambda^2 - 25921}$$

Transmission Region, (External Transmittance ≥10% with 2.0 mm. thickness)   0.25 - 80μ

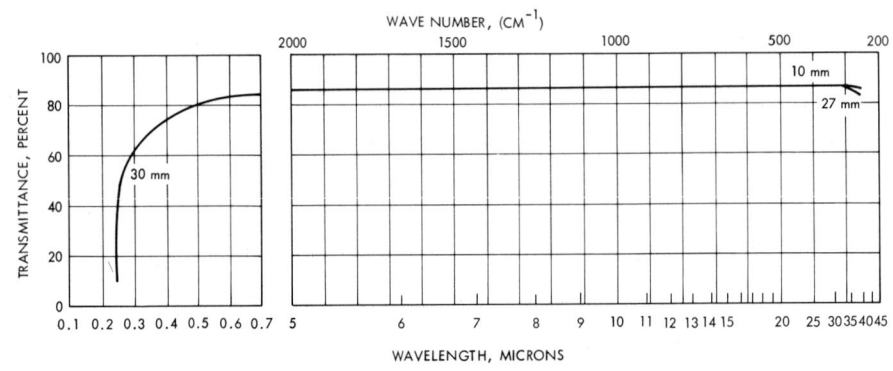

(REF. 1)

MATERIAL: CESIUM IODIDE

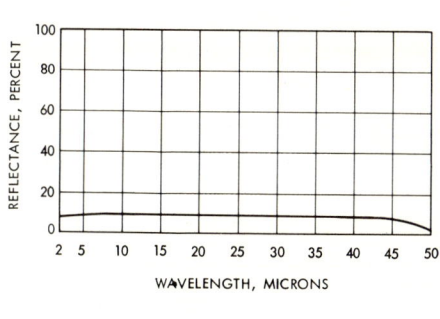

(REF. 2)

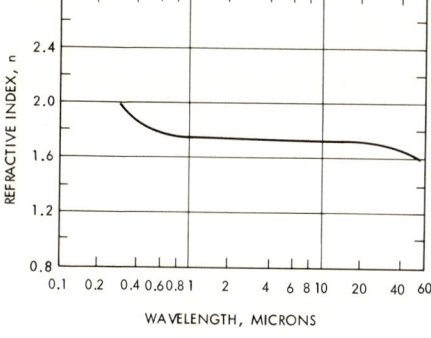

(REF. 3)

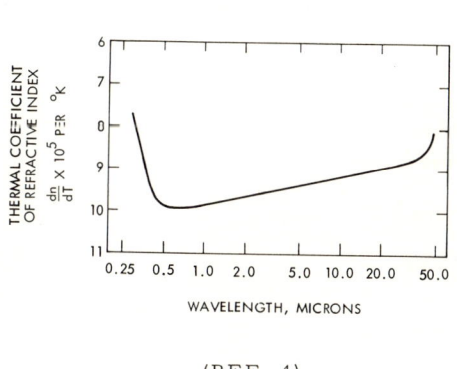

(REF. 4)

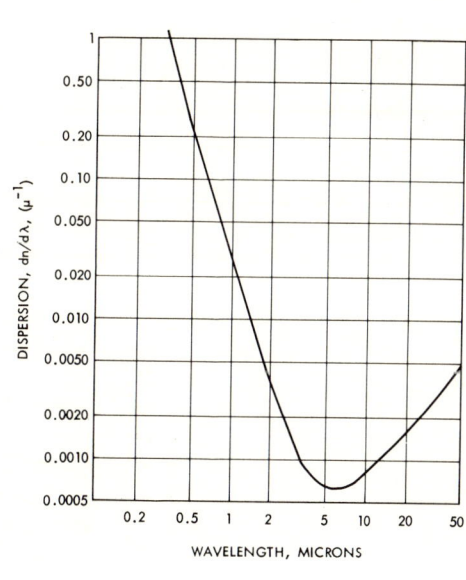

(REF. 4)

REFERENCES:

1. A. Smakula, et al., "Harshaw Optical Crystals", Harshaw Chemical Co., Cleveland, (1967).

2. D. E. McCarthy, Appl. Optics, 2, 591-595, (1963).

3. A. Smakula, Optica Acta, 9, 205-222, (1962).

4. W. S. Rodney, J. Opt. Soc. Am., 45, 987-992, (1955).

COPPER

OPTICAL MATERIALS PROPERTIES: MATERIAL: <u>COPPER</u>
DATA SHEET   <u>(Film)</u>

INTRODUCTION: This data sheet contains optical data for copper films, in addition to thermal, physical, and mechanical properties for bulk copper.

PHYSICAL PROPERTIES, (298°K)

Density, (g/cm^3)   8.94

Melting/Softening Temp. (°K)   1356

Solubility in Water, (g./100 g. H_2O)   <0.01

MECHANICAL PROPERTIES*, (298°K)

Young's Modulus, (psi) 17×10^6

Hardness, (Vickers)   51

THERMAL PROPERTIES, (298°K)

Linear Expansion Coeff., (°K)$^{-1}$ 17.7×10^{-6}

Thermal Conductivity (10^{-4} cal/(cm sec °K)   9400

Specific Heat, (cal/g)/°K   0.092

OPTICAL PROPERTIES, (298°K)

Dispersion Equation   Not available

Transmission Region, (External Transmittance ≥10% with _____ mm. thickness)   Not available

*Annealed bulk metal

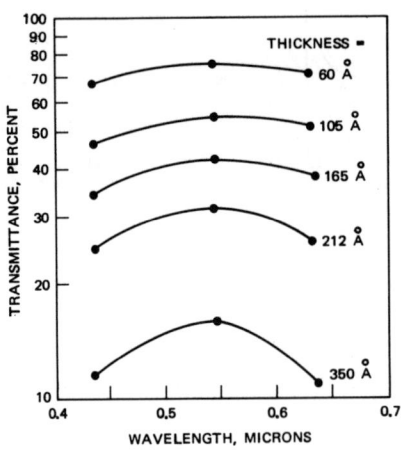

(REF. 1)

MATERIAL: COPPER
(Film)

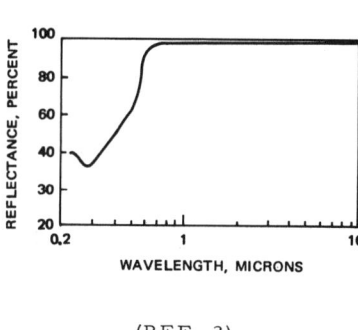

(REF. 2)

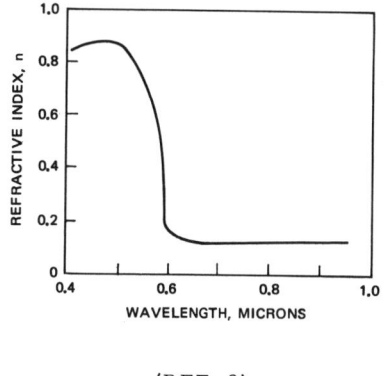

(REF. 3)

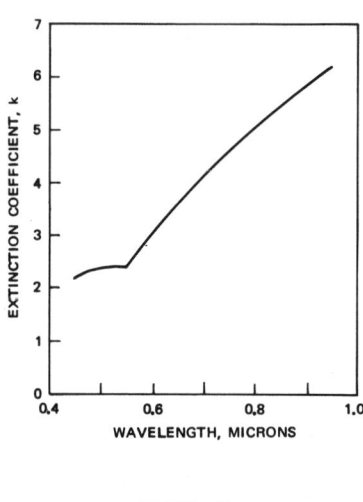

(REF. 3)

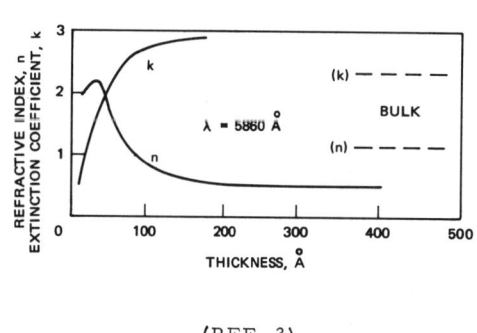

(REF. 3)

REFERENCES:

1. R.S. Adhay, Can. J. Phys. <u>38</u>, 1970-1576, (1960).

2. W.L. Wolfe, Ed., "Handbook of Military Infrared Technology," U.S. Govt. Printing Office, Washington, (1965).

3. O.S. Heavens, "Optical Properties of Thin Solid Films," Academic Press, N.Y., (1955).

CUPROUS CHLORIDE

OPTICAL MATERIALS PROPERTIES
DATA SHEET

MATERIAL: CUPROUS CHLORIDE

INTRODUCTION: This data sheet presents information on single crystal cuprous chloride.

PHYSICAL PROPERTIES, (298°K)
  Density, (g/cm^3) _____ 4.14
  Melting/Softening Temp. (°K) _____ 703
  Solubility in Water, (g./100 g. H_2O) _____ 1.5

MECHANICAL PROPERTIES, (298°K)
  Young's Modulus, (psi) _____ Not available
  Hardness (microhardness) _____ 11 kg/mm^2

THERMAL PROPERTIES, (298°K)
  Linear Expansion Coeff., (°K)$^{-1}$ _____ 13.6×10^{-6}
  Thermal Conductivity (10^{-4} cal/(cm sec °K)) _____ Not available
  Specific Heat, (cal/g)/°K _____ 0.12

OPTICAL PROPERTIES, (298°K)
  Dispersion Equation _____ Not available
  Transmission Region, (External Transmittance ≥10% with 9.1 mm. thickness) _____ 0.4 - 19µ

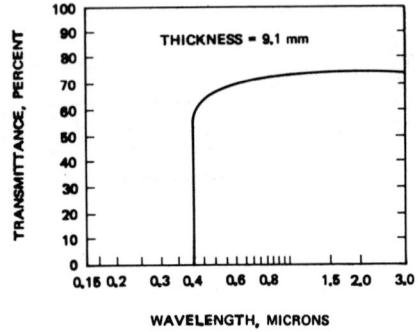

(REF. 1)

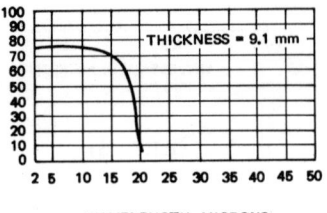

(REF. 2)

MATERIAL: **CUPROUS CHLORIDE**

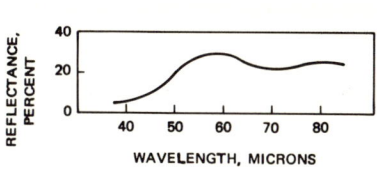

(REF. 3)

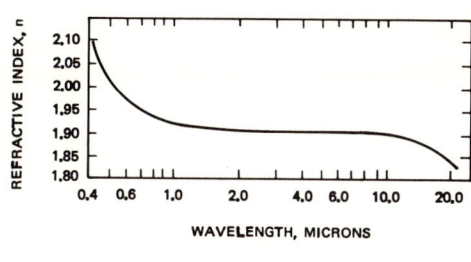

(REF. 4)

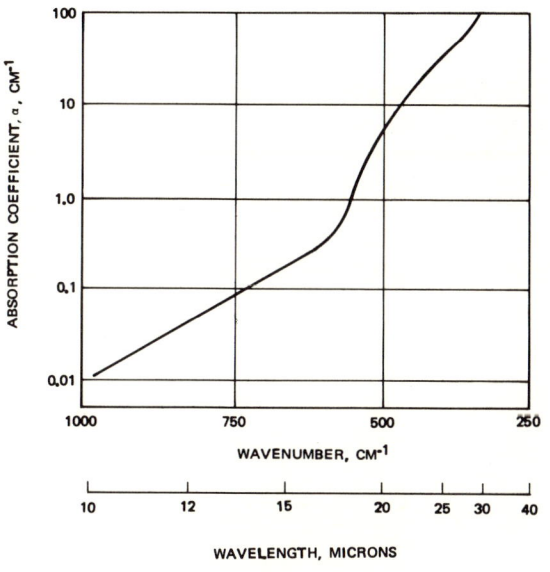

(REF. 5)

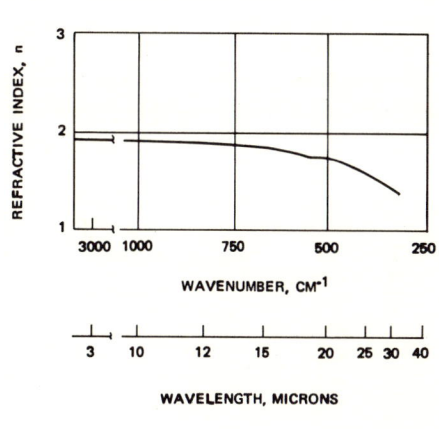

(REF. 5)

REFERENCES:

1. D.E. McCarthy, Appl. Optics, 6, 1896-1898, (1967).
2. D.E. McCarthy, Appl. Optics, 4, 317-320, (1965).
3. J.N. Plendl, et al, Appl. Optics, 5, 397-401, (1966).
4. A. Feldman and D. Horowitz, Opt. Soc. Am., 59, 1406-1408, (1969).
5. A. Smakula, Report No. AD 663 745, (1967).

GALLIUM ARSENIDE

OPTICAL MATERIALS PROPERTIES MATERIAL: __GALLIUM__
DATA SHEET __ARSENIDE__

INTRODUCTION: This data sheet contains information for intrinsic single crystal gallium arsenide.

PHYSICAL PROPERTIES, (298°K)

Density, (g/cm³) __5.307__

Melting/Softening Temp. (°K) __1511__

Solubility in Water, (g./100 g. H₂O) __<0.005__

MECHANICAL PROPERTIES, (298°K)

Young's Modulus, (psi) __not available__

Hardness, (Knoop) __750__

THERMAL PROPERTIES, (298°K)

Linear Expansion Coeff. (°K)⁻¹ __6.8 x 10⁻⁶__

Thermal Conductivity (10⁻⁴ cal/(cm sec °K) __1080__

Specific Heat, (cal/g)/°K __0.076__

OPTICAL PROPERTIES, (298°K)

Dispersion Equation: __not available__

Transmission Region, (External Transmittance ≥ 10% with __2.0__ mm. thickness) __1.0 - 15μ__

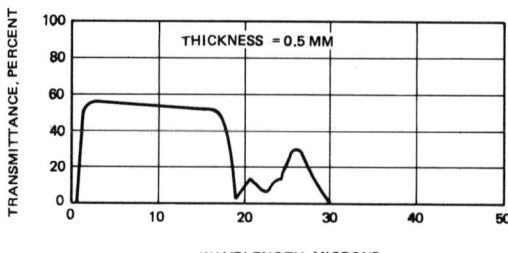

(REF. 1)

MATERIAL: GALLIUM ARSENIDE

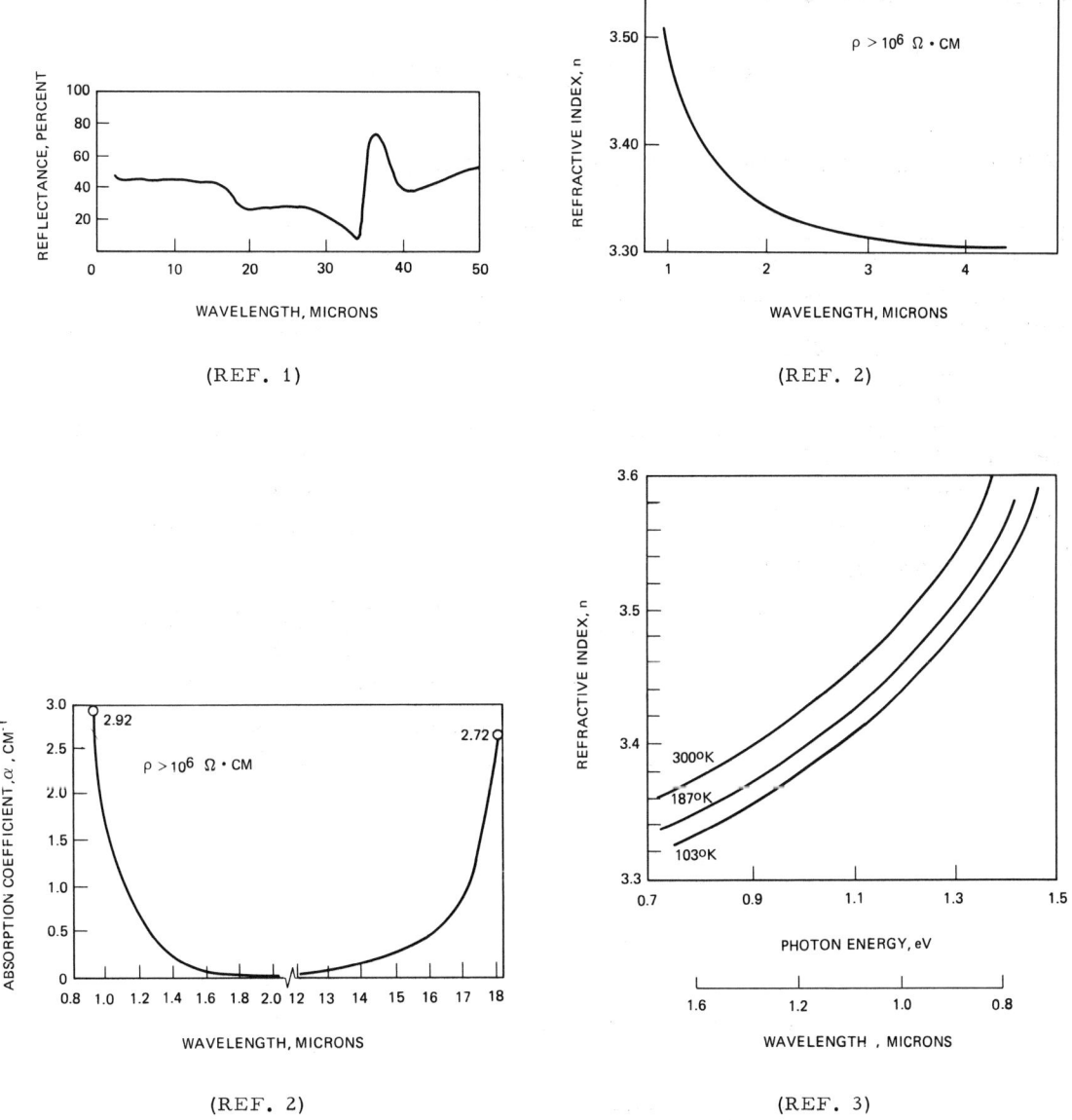

(REF. 1)

(REF. 2)

(REF. 2)

(REF. 3)

REFERENCES:

1. D.E. McCarthy, Appl. Optics, 7, 1997-2000, (1968).

2. E. Hirschmann, and T.E. Walsh, Report No. NASA TN D-4049, (June 1967).

3. D.T.F. Marple, J. Appl. Phys. 35, 1241-1242, (1964).

GERMANIUM

OPTICAL MATERIALS PROPERTIES DATA SHEET   MATERIAL: GERMANIUM

INTRODUCTION: This data sheet summarizes results for intrinsic single crystal germanium.

PHYSICAL PROPERTIES, (298°K)
- Density, (g/cm^3)   5.33
- Melting/Softening Temp. (°K)   1209
- Solubility in Water, (g./100 g. H$_2$O)   <0.005

MECHANICAL PROPERTIES, (298°K)
- Young's Modulus, (psi) 14.9×10^6
- Harness, (Knoop)   700-880

THERMAL PROPERTIES, (298°K)
- Linear Expansion Coeff. (°K)$^{-1}$ 5.5×10^{-6}
- Thermal Conductivity (10^{-4} cal/(cm sec °K)   1400
- Specific Heat, (cal/g)/°K   0.074

OPTICAL PROPERTIES, (298°K)

Dispersion Equation

$$n = A + BL + CL^2 + D\lambda^2 + E\lambda^2$$

where

A = 3.99931; B = 0.391707; C = 0.163492; D = -0.0000060; E = 0.000000053
(valid between 2.0 and 13.5µ)

Transmission Region, (External Transmittance ≥10% with 2.0 mm. thickness) 1.8 - 23µ

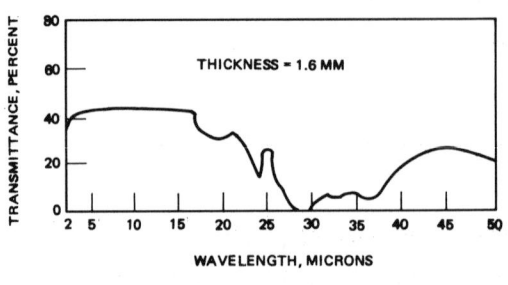

(REF. 1)

MATERIAL: GERMANIUM

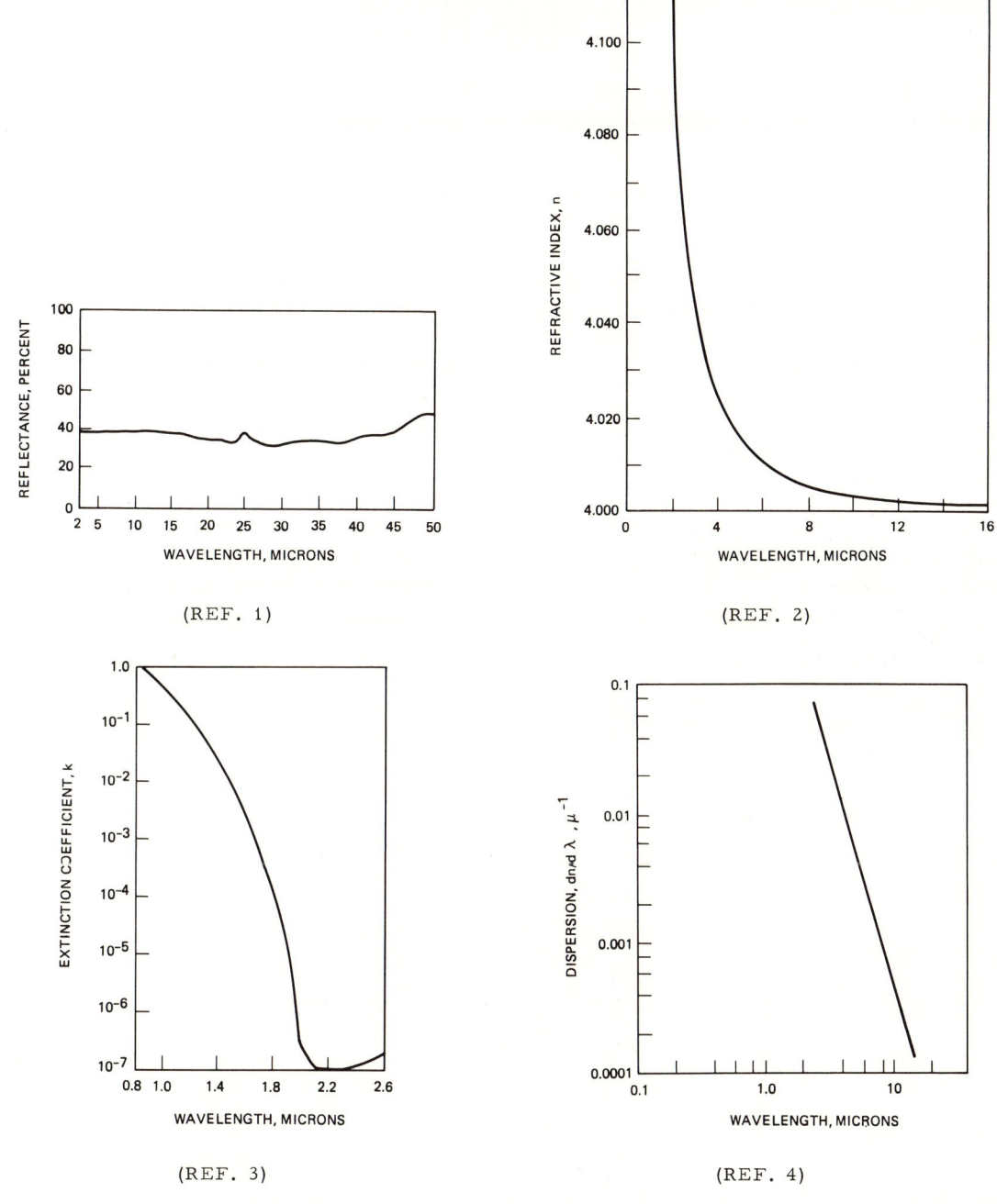

REFERENCES:

1. D. E. McCarthy, Appl. Optics, 2, 591-595, (1963).

2. C. D. Salzberg and J. J. Villa, J. Opt. Soc. Am., 47, 244-246, (1957).

3. H. B. Briggs, J. Opt. Soc. Am., 42, 686-687, (1952).

4. W. L. Wolfe, Ed., "Handbook of Military Infrared Technology", U. S. Govt. Printing Office, Washington, (1965).

GOLD

OPTICAL MATERIALS PROPERTIES　　　　MATERIAL: GOLD
DATA SHEET　　　　　　　　　　　　　　　　(FILM)

INTRODUCTION: This data sheet presents information for gold film, formed by evaporation. The optical data may be compromised by uncertainty concerning the crystallinity of the film.

PHYSICAL PROPERTIES, (298°K)
- Density, (g/cm^3) _____ 19.32
- Melting/Softening Temp. (°K) _____ 1336
- Solubility in Water, (g./100 g. H$_2$O) _____ <0.001

MECHANICAL PROPERTIES, (298°K)
- Young's Modulus, (psi) _____ 0.78 x 10^6
- Hardness, (Vickers) _____ 50-120

THERMAL PROPERTIES, (298°K)
- Linear Expansion Coeff. (°K)$^{-1}$ _____ 14.1 x 10^{-6}
- Thermal Conductivity (10^{-4} cal/(cm sec °K) _____ 7100
- Specific Heat, (cal/g)/°K _____ 0.031

OPTICAL PROPERTIES, (298°K)
- Dispersion Equation: not available
- Transmission Region, (External Transmittance ≥ 10% with _____ mm. thickness) _____ not available

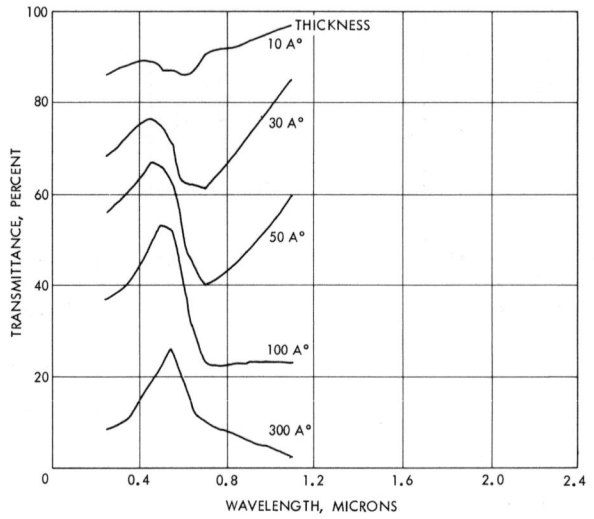

(REF. 1)

MATERIAL: __GOLD__
          __(FILM)__

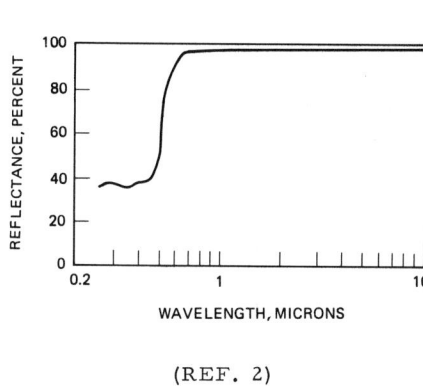

(REF. 2)

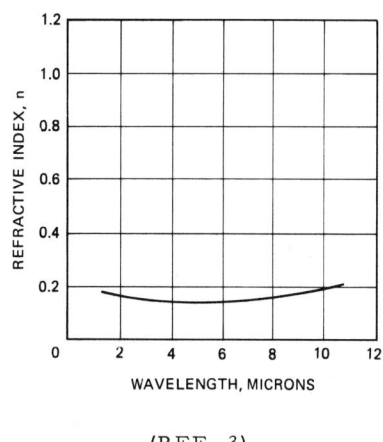

(REF. 3)

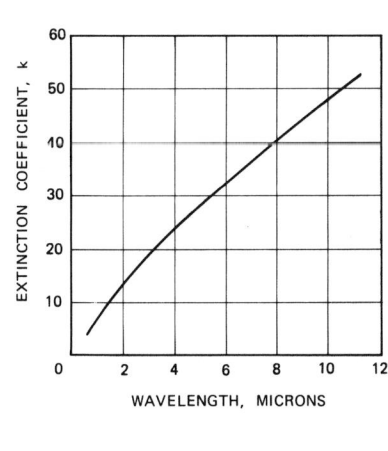

(REF. 3)

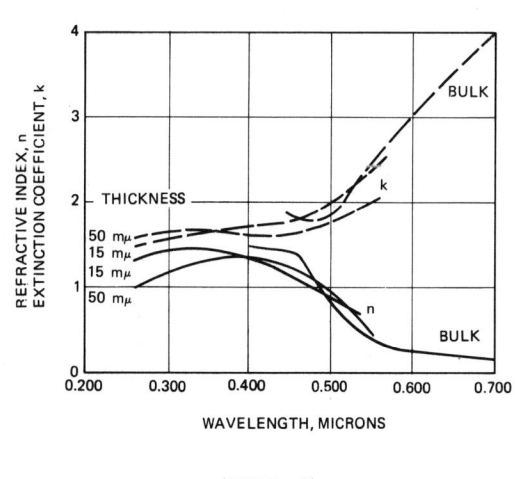

(REF. 3)

REFERENCES:

1. F. Goos, Z. Phys., __106__, 606-619, (1937).

2. W.L. Wolfe, Ed., "Handbook of Military Infrared Technology", U.S. Govt. Printing Office, Washington, (1965).

3. O.S. Heavens, in "Reports on Progress in Physics", A.C. Sickland, Ed., __23__, 30-65, (1960).

INDIUM ARSENIDE

OPTICAL MATERIALS PROPERTIES DATA SHEET   MATERIAL: INDIUM ARSENIDE

INTRODUCTION: This data sheet contains information for single crystal indium arsenide.

PHYSICAL PROPERTIES, (298°K)
Density, (g/cm^3) 5.66
Melting/Softening Temp. (°K) 1215
Solubility in Water, (g./100 g. H$_2$O) <0.05

MECHANICAL PROPERTIES, (298°K)
Young's Modulus, (psi) Not available
Hardness, (Knoop) 380

THERMAL PROPERTIES, (298°K)
Linear Expansion Coeff., (°K)$^{-1}$ 5.3×10^{-6}
Thermal Conductivity (10^{-4} cal/(cm sec °K)) 640 (polycrystalline)
Specific Heat, (cal/g)/°K 0.061

OPTICAL PROPERTIES, (298°K)
Dispersion Equation Not available
Transmission Region, (External Transmittance ≥10% with 2.0 mm. thickness) 3.8 - 7.0μ

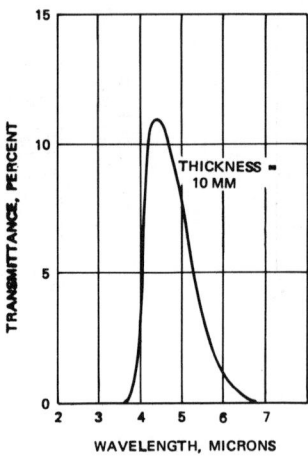

(REF. 1)

MATERIAL: INDIUM ARSENIDE

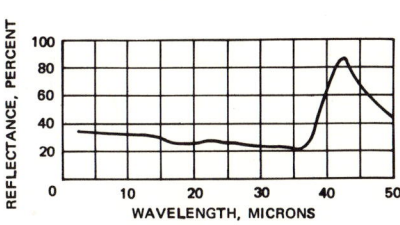

(REF. 2)

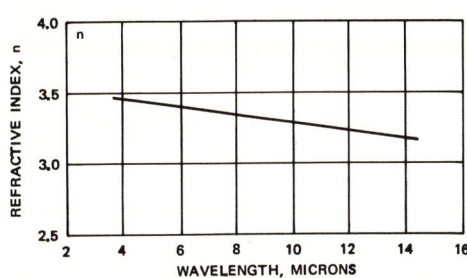

(REF. 1)

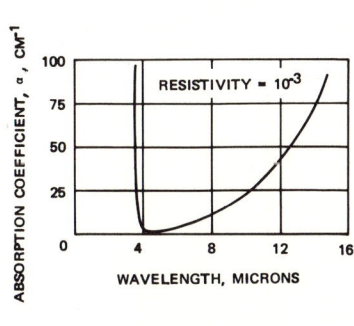

(REF. 3)

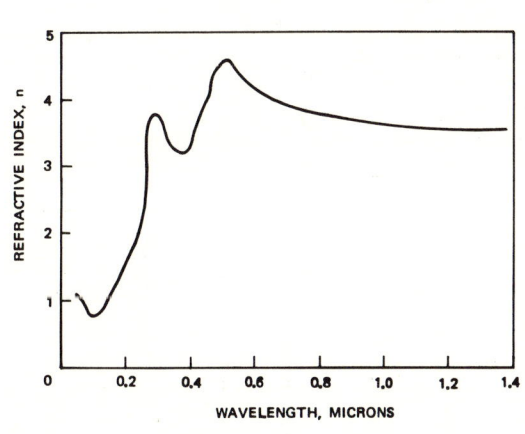

(REF. 4)

REFERENCES:

1. P. Billard, Acta Electronica, 6, 75-169, (1962).

2. D.E. McCarthy, Appl. Optics, 7, 1997-2000, (1968).

3. F. Oswald and R. Schade, Z. Naturforsch, 9a, 611-617, (1954).

4. B.O. Seraphin and H.E. Bennett in "Semiconductors and Semimetals," R. K. Willard and R.C. Beer, Eds, Vol. 3, Academic Press, N.Y., (1967).

LEAD SULFIDE

OPTICAL MATERIALS PROPERTIES DATA SHEET

MATERIAL: LEAD SULFIDE

INTRODUCTION: This data sheet furnishes information on lead sulfide with optical data applying to single crystal lead sulfide.

PHYSICAL PROPERTIES, (298°K)
- Density, (g/cm^3) _____ 7.5
- Melting/Softening Temp. (°K) _____ 1392
- Solubility in Water, (g./100 g. H$_2$O) _____ 3×10^{-6}

MECHANICAL PROPERTIES, (298°K)
- Young's Modulus, (psi) _____ 1.64×10^{-6}
- Hardness, (Mohs) _____ 2.5

THERMAL PROPERTIES, (298°K)
- Linear Expansion Coeff. (°K)$^{-1}$ _____ 24×10^{-6}
- Thermal Conductivity (10^{-4} cal/(cm sec °K)) _____ 500
- Specific Heat, (cal/g)/°K _____ 0.039

OPTICAL PROPERTIES, (298°K)
- Dispersion Equation: not available
- Transmission Region, (External Transmittance <10% with 2.0 mm. thickness) 3.0 - 7.0µ

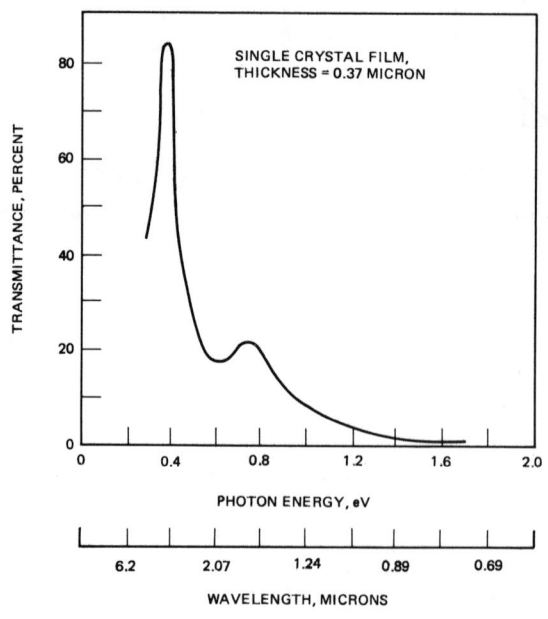

(REF. 1)

MATERIAL: LEAD SULFIDE

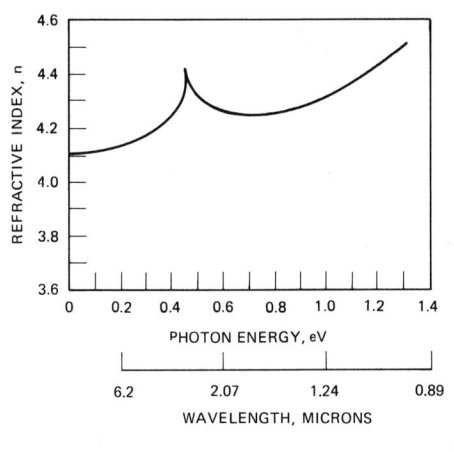

(REF. 1)

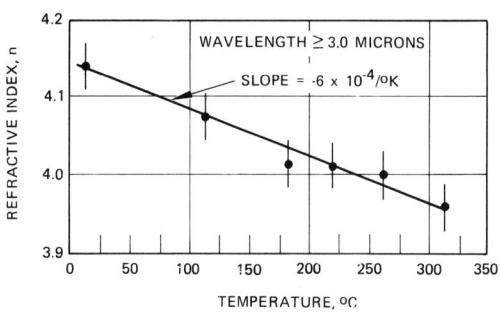

(REF. 3)

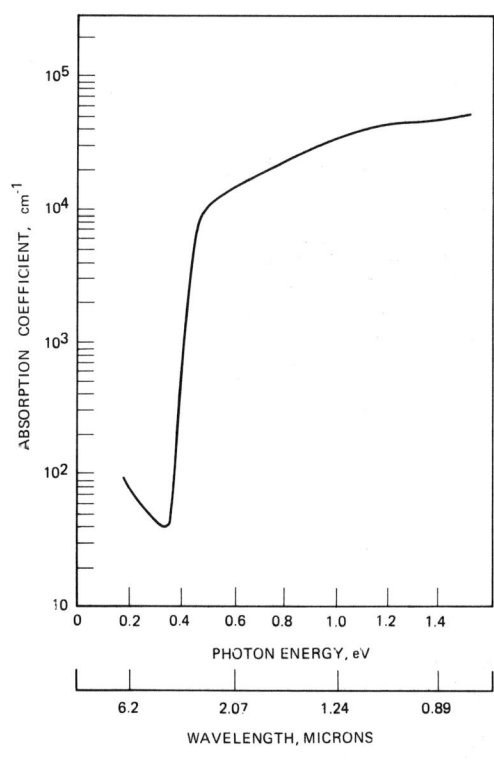

(REF. 1)

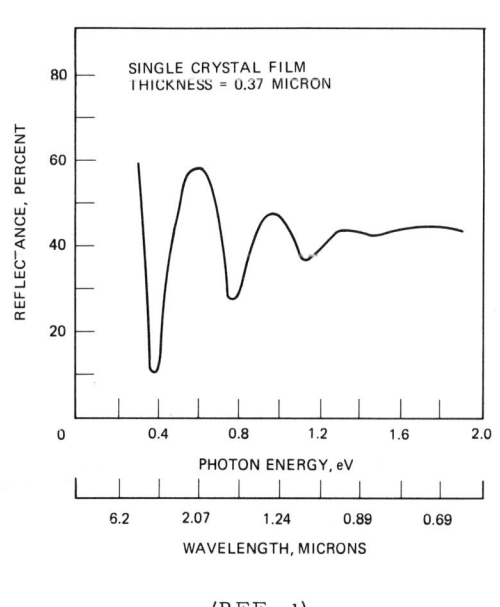

(REF. 1)

REFERENCES:

1. R.B. Schoolar and J.R. Dixon, Phys. Rev., 137, A667-A670, (1965).

2. W.W. Scanlon, Phys. Rev., 109, 47-50, (1958).

3. D.G. Avery, Proc. Phys. Soc. (London), B67, 2-8, (1954).

LITHIUM FLUORIDE

OPTICAL MATERIALS PROPERTIES
DATA SHEET

MATERIAL: <u>LITHIUM FLUORIDE</u>

INTRODUCTION: <u>This data sheet presents information for single crystal lithium fluoride.</u>

PHYSICAL PROPERTIES, (298°K)
- Density, (g/cm^3) <u>2.64</u>
- Melting/Softening Temp. (°K) <u>1403</u>
- Solubility in Water, (g./100 g. H$_2$O) <u>0.27 (291°K)</u>

MECHANICAL PROPERTIES, (298°K)
- Young's Modulus, (psi) <u>9.4 x 10^6</u>
- Hardness, (Knoop) <u>~113 (600 g.)</u>

THERMAL PROPERTIES, (298°K)
- Linear Expansion Coeff. (°K)$^{-1}$ <u>37 x 10^{-6}</u>
- Thermal Conductivity (10^{-4} cal/(cm sec °K) <u>340</u>
- Specific Heat, (cal/g)/°K <u>0.373 (283°K)</u>

OPTICAL PROPERTIES, (298°K)

Dispersion Equation:

$$n = A + BL + CL^2 + D\lambda^2 + E\lambda^4$$

where

A = 1.38761; B = 0.001796; C = -0.000041;
D = -0.0023045; E = -0.00000557
(between 0.5 and 6.0μ).

Transmission Region, (External Transmittance ≥ 10% with <u>2.0</u> mm. thickness) <u>0.12 - 9.0μ</u>

(REF. 1)

MATERIAL: **LITHIUM FLUORIDE**

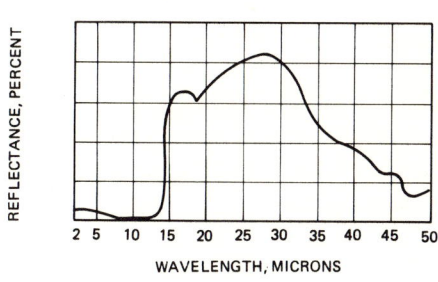

(REF. 2)

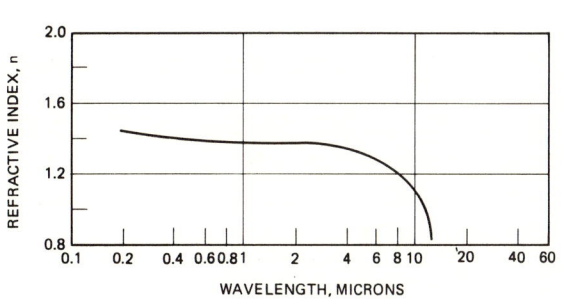

(REF. 3)

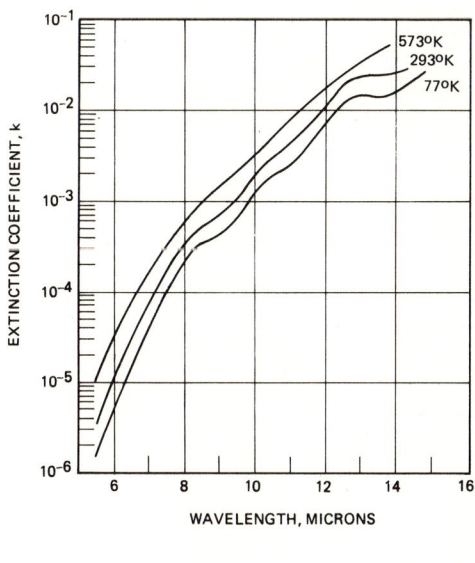

(REF. 4)

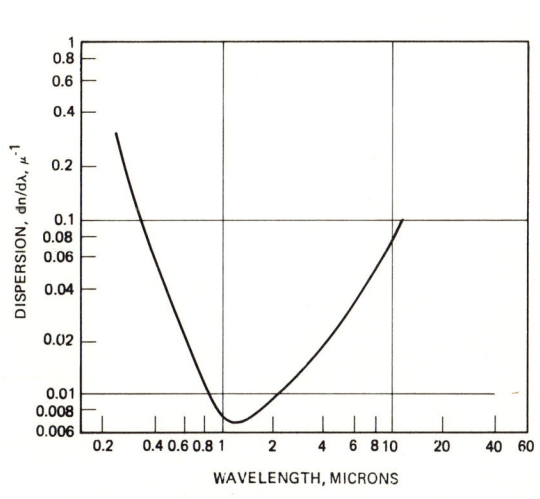

(REF. 3)

REFERENCES:

1. A. Smakula, "Harshaw Optical Crystals", The Harshaw Chemical Co., Cleveland, (1967).

2. D.E. McCarthy, Appl. Optics, 2, 591-595, (1963).

3. A. Smakula, Optica Acta, 9, 205-222, (1962).

4. M. Klier, Z. Physik, 150, 49-63, (1958).

LITHIUM NIOBATE

OPTICAL MATERIALS PROPERTIES
DATA SHEET

MATERIAL: <u>LITHIUM NIOBATE</u>

INTRODUCTION: <u>This data sheet contains information for single crystal lithium niobate.</u>

PHYSICAL PROPERTIES, (298°K)
- Density, (g/cm^3) <u>   4.70   </u>
- Melting/Softening Temp. (°K) <u>   1533   </u>
- Solubility in Water, (g./100 g. H$_2$O) <u>   <0.005   </u>

MECHANICAL PROPERTIES, (298°K)
- Young's Modulus, (psi) <u> not available </u>
- Hardness, (Mohs) <u>   ~5   </u>

THERMAL PROPERTIES, (298°K)
- Linear Expansion Coeff. (°K)$^{-1}$ <u>   16.7 x 10^{-6} </u>
- Thermal Conductivity (10^{-4} cal/(cm sec °K) <u>   100   </u>
- Specific Heat, (cal/g)/°K <u>   0.153   </u>

OPTICAL PROPERTIES, (298°K)
- Dispersion Equation: <u> not available </u>
- Transmission Region, (External Transmittance ≥ 10% with <u> 10 </u> mm. thickness) <u>   0.33 - 5.2µ   </u>

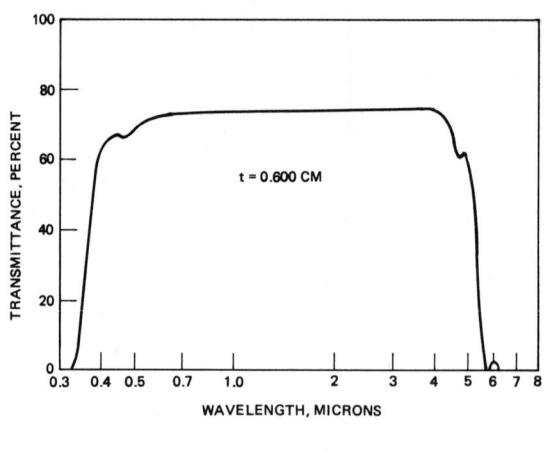

(REF. 1)

MATERIAL: LITHIUM NIOBATE

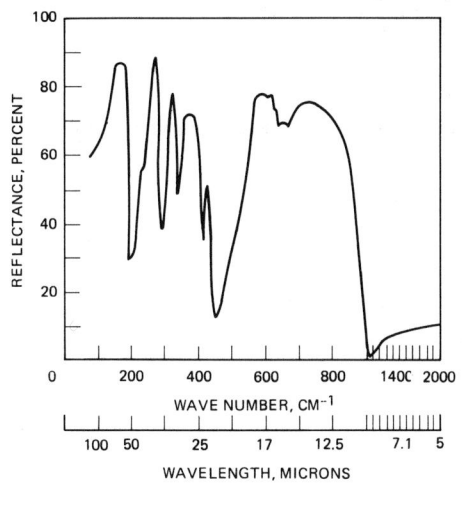

(REF. 2)

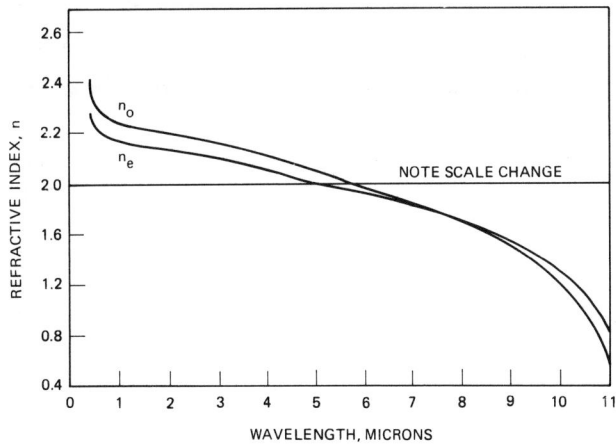

(REF. 3)

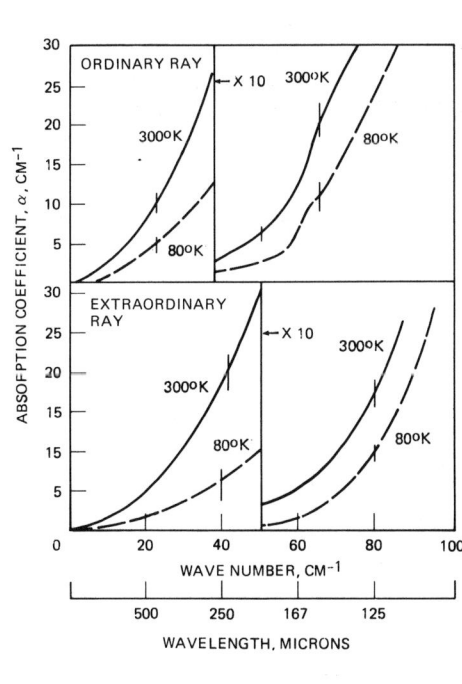

(REF. 4)

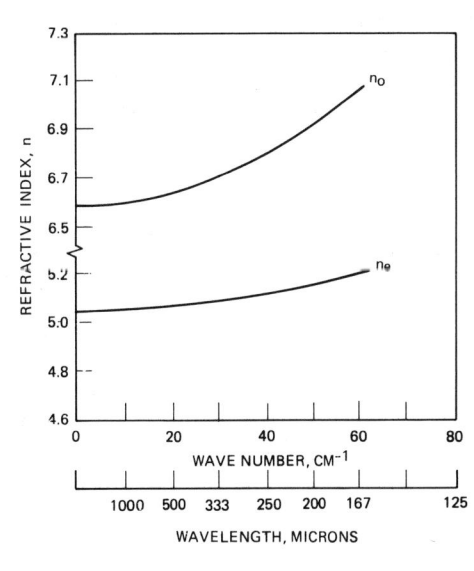

(REF. 4)

REFERENCES:

1. K. Nassau, et al., J. Phys. Chem. Solids, 27, 989-996, (1966).
2. J.D. Axe and D.F. O'Kane, Appl. Phys. Letters, 9, 58-60, (1966).
3. A.S. Barker, Jr. and R. Loudon, Phys. Rev. 158, 433-445, (1967).
4. D.R. Bosomworth, Appl. Phys. Letters, 9, 330-331, (1966).

MAGNESIUM FLUORIDE
(Film)

OPTICAL MATERIALS PROPERTIES
DATA SHEET

MATERIAL: MAGNESIUM FLUORIDE (Film)

INTRODUCTION: This data sheet applies to freshly evaporated films of magnesium fluoride deposited on glass.

PHYSICAL PROPERTIES, (298°K)
- Density, (g/cm^3)   ~3.2
- Melting/Softening Temp. (°K)   1528
- Solubility in Water, (g./100 g. H$_2$O)   Not available

MECHANICAL PROPERTIES, (298°K)
- Young's Modulus, (psi)   Not available
- Hardness, Knoop)   Not available

THERMAL PROPERTIES, (298°K)
- Linear Expansion Coeff., (°K)$^{-1}$ Not available
- Thermal Conductivity (10^{-4} cal/(cm sec °K)   Not available
- Specific Heat, (cal/g)/°K   Not available

OPTICAL PROPERTIES, (298°K)
- Dispersion Equation   Not available
- Transmission Region, (External Transmittance ≥10% with ~1/4 λ thickness)   0.2 - 5.0μ

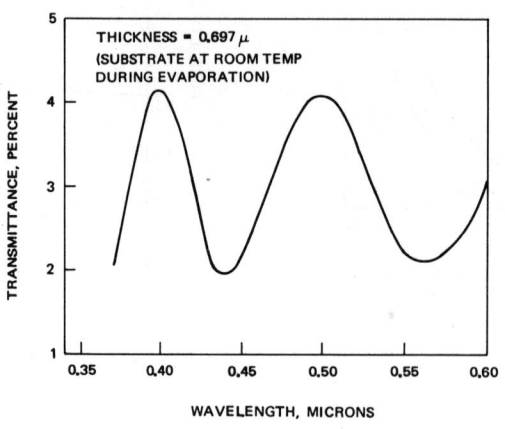

(REF. 1)

MATERIAL: MAGNESIUM FLUORIDE
(Film)

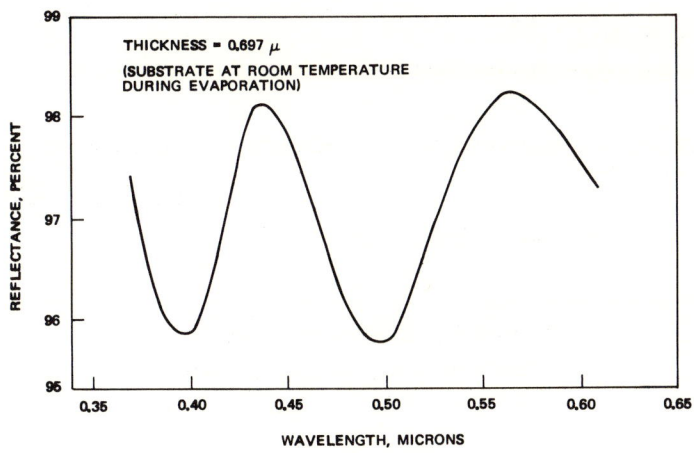

(REF. 1)

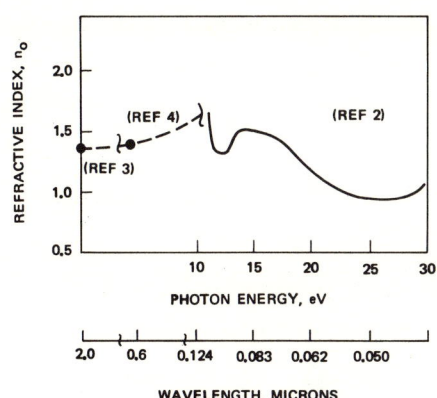

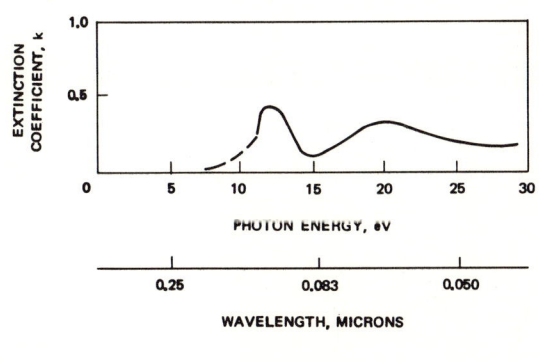

(REF. 2)

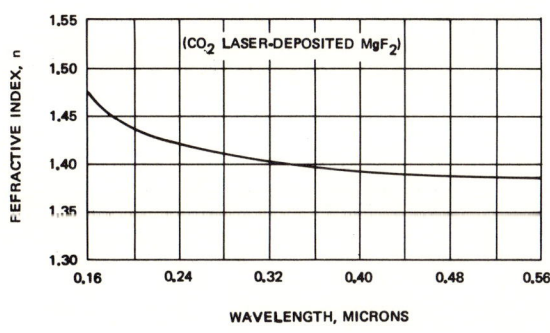

(REF. 5)

REFERENCES:

1. N. Morita, J. Phys. Soc., Japan, 11, 975-980, (1956).

2. M.W. Williams, et al, J. Appl. Phys., 38, 1701-1705, (1967).

3. J.R. Jenness, Jr., J. Am. Opt. Soc,, 46, 157-159, (1956).

4. J.F. Hall, Jr., and W.F.C. Ferguson, J. Am. Opt. Soc., 45, 74-75, (1955).

5. G. Hass and J.B. Ramsey, Appl. Optics, 8, 1115-1118, (1969).

MAGNESIUM FLUORIDE
(Single Crystal)

OPTICAL MATERIALS PROPERTIES DATA SHEET

MATERIAL: MAGNESIUM FLUORIDE (Single crystal)

INTRODUCTION: This data sheet contains information for single crystal magnesium fluoride

PHYSICAL PROPERTIES, (298°K)
- Density, (g/cm³)   3.177
- Melting/Softening Temp. (°K)   1498
- Solubility in Water, (g./100 g. H₂O)   0.0076 (291°K)

MECHANICAL PROPERTIES, (298°K)
- Young's Modulus, (psi)   not available
- Hardness, (Mohs)   415 (100 g.)

THERMAL PROPERTIES, (298°K)
- Linear Expansion Coeff. (°K)⁻¹   4 x 10⁻⁶
- Thermal Conductivity (10⁻⁴ cal/(cm sec °K)   not available
- Specific Heat, (cal/g)/°K   0.284

OPTICAL PROPERTIES, (298°K)

Dispersion Equation:

$n_o = 1.36957 + 0.0035821/[\lambda - 0.14925]$

$n_e = 1.38100 + 0.0037415/[\lambda - 0.14947]$

where

λ = wavelength between 0.4 and 0.7 microns

Transmission Region, (External Transmittance ≥10% with 2.0 mm. thickness)   0.1 - 9.7µ

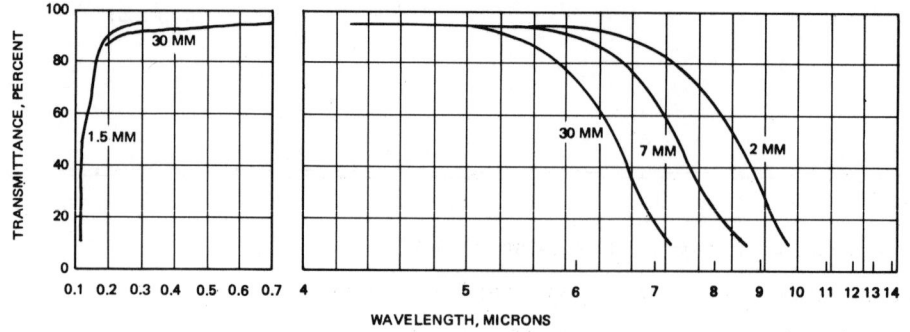

(REF. 1)

MATERIAL: **MAGNESIUM FLUORIDE** (Single crystal)

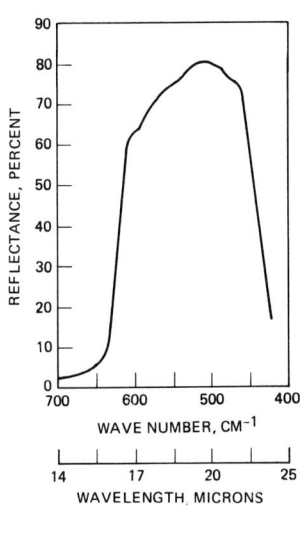

(REF. 2)

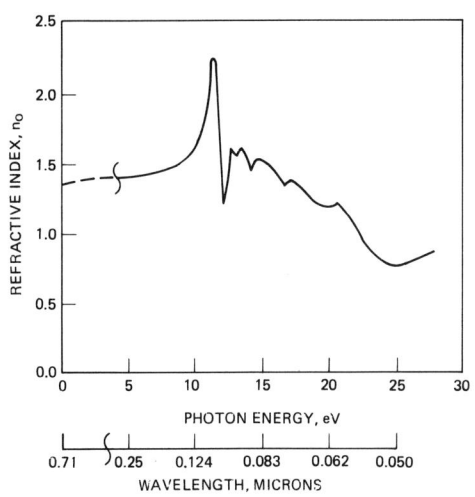

(REF. 2, 3)

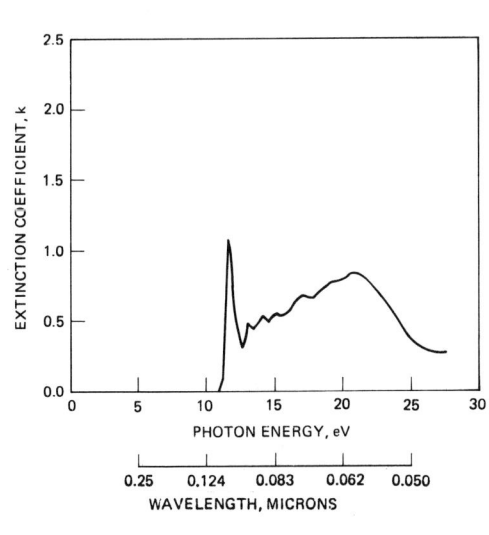

(REF.)

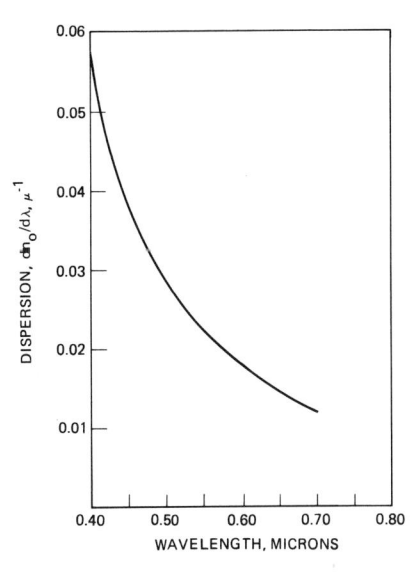

(REF. 2)

REFERENCES:

1. A. Smakula, et al., "Harshaw Optical Crystals", The Harshaw Chemical Co., Cleveland, (1967).

2. A. Duncanson and R.W. Stevenson, Proc. Royal Soc. London, 72, 1001-1006, (1958).

3. M.W. Williams, et al., J. Appl. Phys., 38, 1701-1705, (1967).

MAGNESIUM OXIDE

OPTICAL MATERIALS PROPERTIES DATA SHEET MATERIAL: MAGNESIUM OXIDE

INTRODUCTION: This data sheet contains information for single crystal magnesium oxide.

PHYSICAL PROPERTIES, (298°K)
- Density, (g/cm^3) __3.58__
- Melting/Softening Temp. (°K) __3073__
- Solubility in Water, (g./100 g. H$_2$O) __1.2×10^{-5} (293°K)__

MECHANICAL PROPERTIES, (298°K)
- Young's Modulus, (psi) __48.2×10^6__
- Hardness, (Knoop) __692 (600 g.)__

THERMAL PROPERTIES, (298°K)
- Linear Expansion Coeff. (°K)$^{-1}$ __10.5×10^{-6}__
- Thermal Conductivity (10^{-4} cal/(cm sec °K) __600__
- Specific Heat, (cal/g)/°K __0.222__

OPTICAL PROPERTIES, (298°K)

Dispersion Equation:

$$n^2 = 2.956362 - 0.1062387\lambda^2 - 0.0000204968\lambda^4 - \frac{0.02195770}{\lambda^2 - 0.01428322}$$

Transmission Region, (External Transmittance $\geq 10\%$ with __2.0__ mm. thickness __$0.25 - 8.5\mu$__

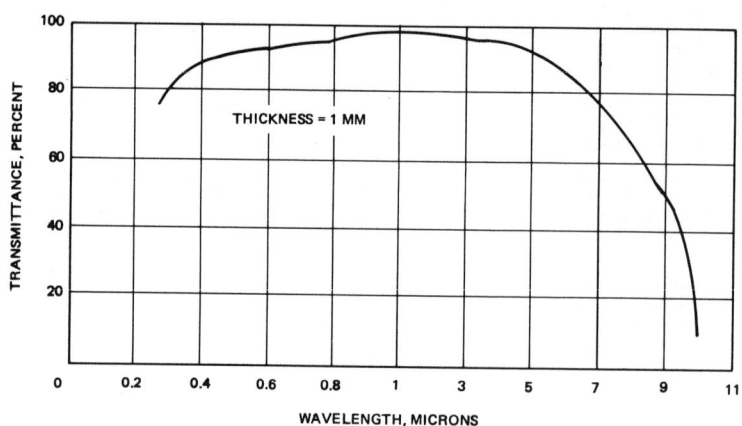

(REF. 1)

MATERIAL: MAGNESIUM OXIDE

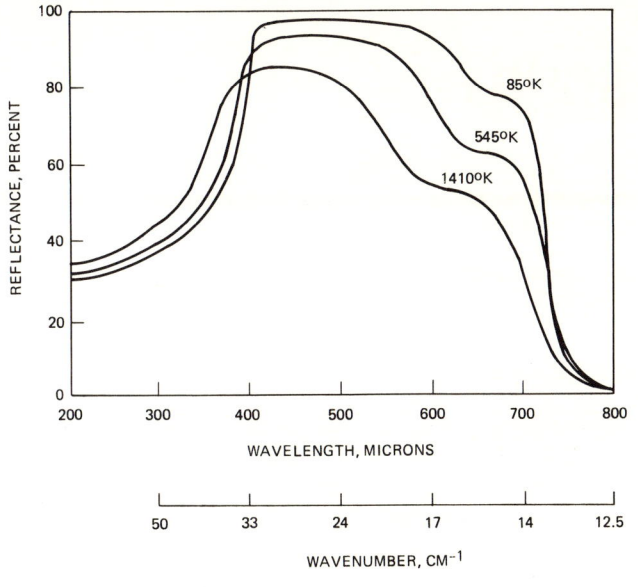

(REF. 2)

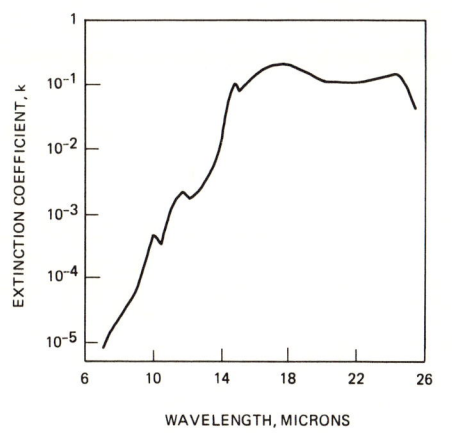

(REF. 4)

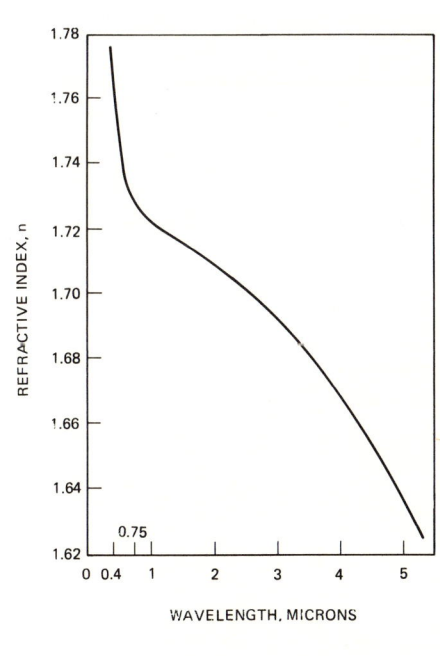

(REF. 3)

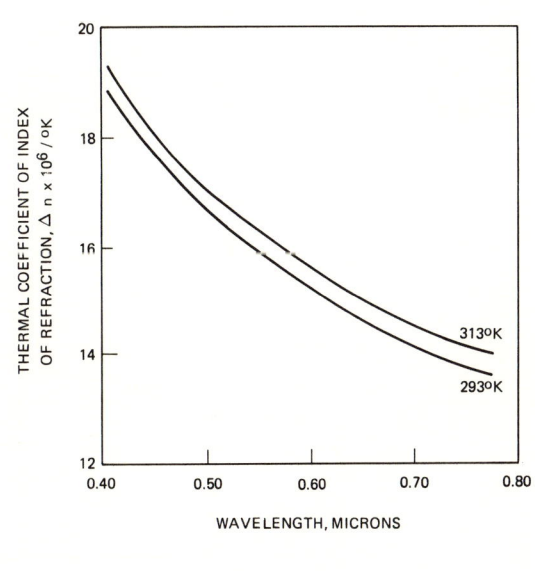

(REF. 3)

REFERENCES:

1. Data Sheet, Semi-Elements, Inc., Saxonburg, Pa.

2. A. Kahan, Report No. AFCRL-66-327, (1966).

3. R.E. Stephens and I.H. Malitson, J. Res. NBS, 49, 249-252, (1952).

4. J.C. Wilmott, Proc. Phys. Soc. London, 63A, 389-402, (1950).

PALLADIUM

OPTICAL MATERIALS PROPERTIES
DATA SHEET

MATERIAL: PALLADIUM
(Film)

INTRODUCTION: This data sheet contains optical data for palladium films as well as thermal, physical and mechanical properties for bulk palladium.

PHYSICAL PROPERTIES, (298°K)

- Density, (g/cm^3) 12.02
- Melting/Softening Temp. (°K) 1827
- Solubility in Water, (g./100 g. H$_2$O) <0.01

MECHANICAL PROPERTIES*, (298°K)

- Young's Modulus, (psi) 1.6×10^7
- Hardness, (Vickers) 37

THERMAL PROPERTIES, (298°K)

- Linear Expansion Coeff., (°K)$^{-1}$ 11.67×10^{-6}
- Thermal Conductivity (10^{-4} cal/(cm sec °K) 1680
- Specific Heat, (cal/g)/°K 0.0584

OPTICAL PROPERTIES, (298°K)

- Dispersion Equation Not available
- Transmission Region, (External Transmittance ≥10% with _____ mm. thickness) Not available

*Annealed bulk metal

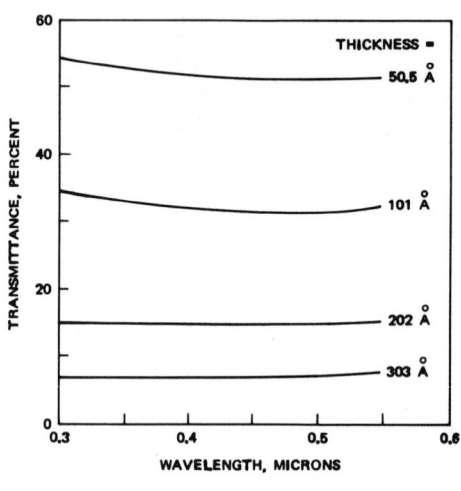

(REF. 1)

MATERIAL: <u>PALLADIUM</u>
<u>(Film)</u>

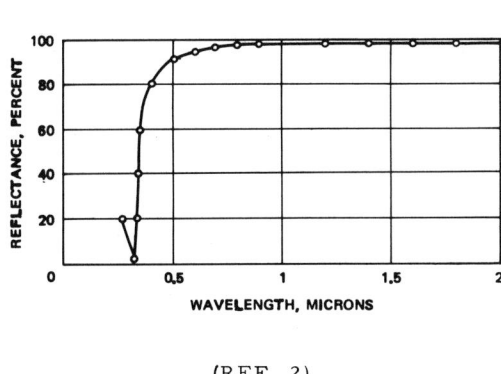

(REF. 2)

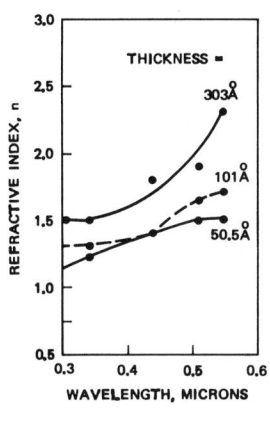

(REF. 1)

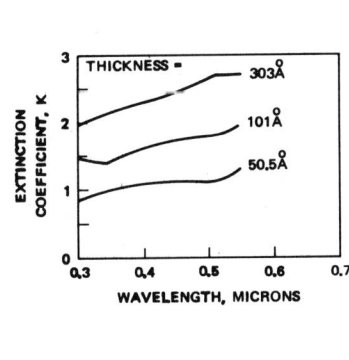

(REF. 1)

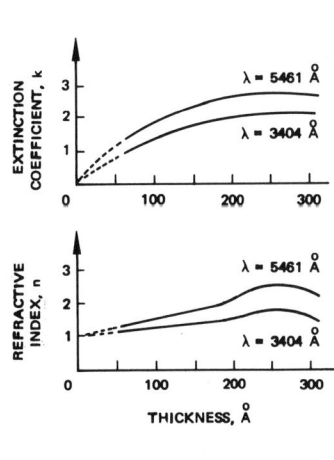

(REF. 1)

REFERENCES:

1. D. Malé and J. Trompette, J. Phys. Radium, <u>18</u>, 128-130, (1957).

2. Degussa, "Edelmetall - Taschenbuch," Degussa, Frankfurt/Main, (1967).

PLATINUM

OPTICAL MATERIALS PROPERTIES DATA SHEET   MATERIAL: **PLATINUM** (Film)

INTRODUCTION: This data sheet contains optical data for platinum film as well as thermal, physical, and mechanical properties for bulk platinum.

PHYSICAL PROPERTIES, (298°K)
- Density, (g/cm^3)   21.45
- Melting/Softening Temp. (°K)   2046
- Solubility in Water, (g./100 g. H$_2$O)   <0.01

MECHANICAL PROPERTIES*, (298°K)
- Young's Modulus, (psi) 2.1×10^7
- Hardness, (Vickers)   37

THERMAL PROPERTIES, (298°K)
- Linear Expansion Coeff., (°K)$^{-1}$ 8.9×10^{-6}
- Thermal Conductivity (10^{-4} cal/(cm sec °K))   1660
- Specific Heat, (cal/g)/°K   0.032

OPTICAL PROPERTIES, (298°K)
- Dispersion Equation   Not available
- Transmission Region, (External Transmittance ≥10% with 2.0 mm. thickness)   Not available

*Annealed bulk metal

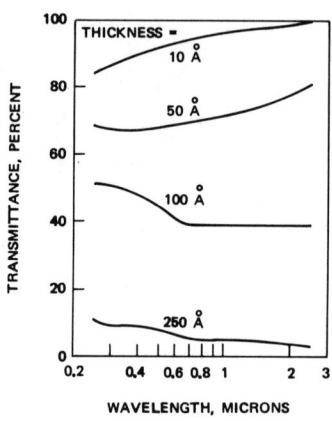

(REF. 1)

MATERIAL: <u>PLATINUM</u>
<u>(Film)</u>

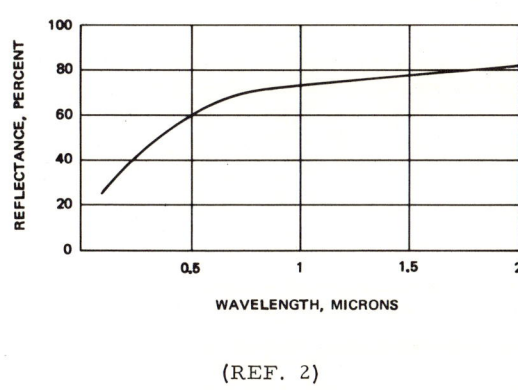

(REF. 2)

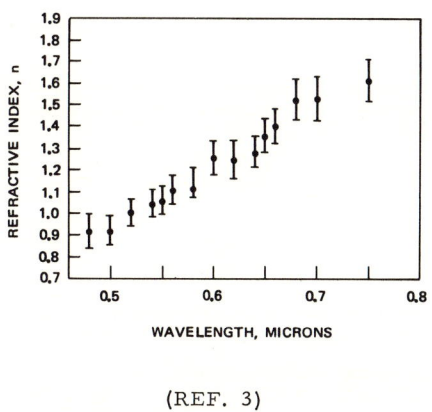

(REF. 3)

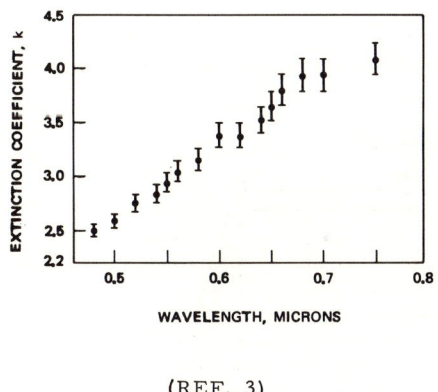

(REF. 3)

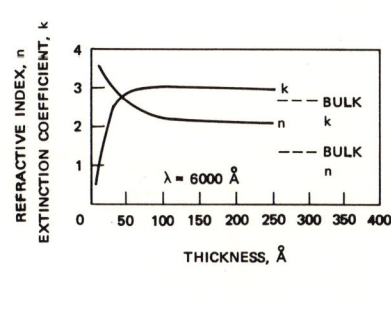

(REF. 4)

REFERENCES:

1. E. Schuch, Ann. der Physik, <u>13</u>, 297-314, (1932).

2. Degussa, "Edelmetall-Taschenbuch," Degussa, Frankfurt/Main, (1967).

3. V. L. Rideout and S. H. Wemple, J. Opt. Soc. Am., <u>56</u>, 749-751, (1966).

4. O. S. Heavens, "Optical Properties of Thin Solid Films," Academic Press, N.Y., (1955).

POTASSIUM BROMIDE

OPTICAL MATERIALS PROPERTIES DATA SHEET   MATERIAL: POTASSIUM BROMIDE

INTRODUCTION: This data sheet contains information for single crystal potassium bromide.

PHYSICAL PROPERTIES, (298°K)

- Density, (g/cm^3)   2.75
- Melting/Softening Temp. (°K)   1003
- Solubility in Water, (g./100 g. H$_2$O)   65.2 (293°K)

MECHANICAL PROPERTIES, (298°K)

- Young's Modulus, (psi)   3.9 x 10^{-6}
- Hardness, (Knoop)   7.0 <100> (200g.)

THERMAL PROPERTIES, (298°K)

- Linear Expansion Coeff. (°K^{-1})   43 x 10^{-6}
- Thermal Conductivity (10^{-4}cal/(cm sec °K)   115 (319°K)
- Specific Heat, (cal/g)/°K   0.104 (273°K)

OPTICAL PROPERTIES, (298°K)

Dispersion Equation:

$$n^2 = 2.3618102 - 0.00058072\lambda^2 + \frac{0.02305269}{\lambda^2 - 0.0245381}$$

(between 0.4 and 0.71μ)

Transmission Region, (External Transmittance ≥10% with __2.0__ mm. thickness)   0.25 - 35μ

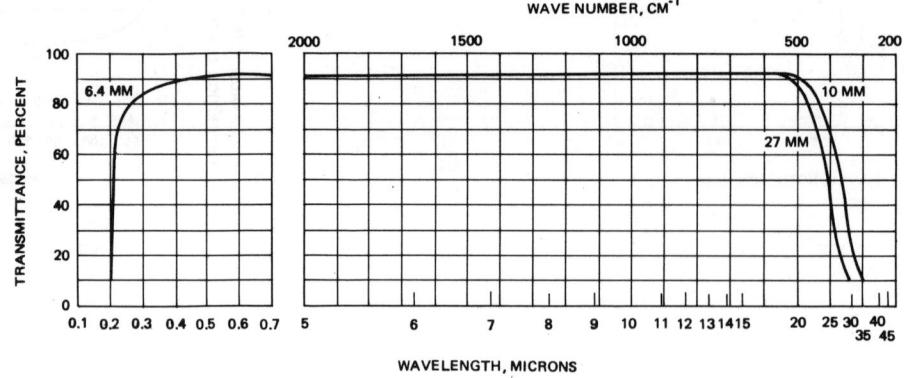

(REF. 1)

MATERIAL: POTASSIUM BROMIDE

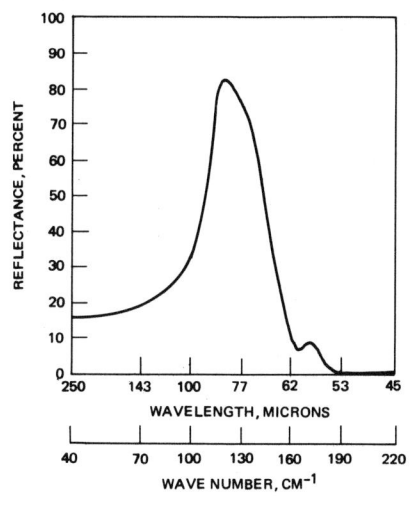

(REF. 2)

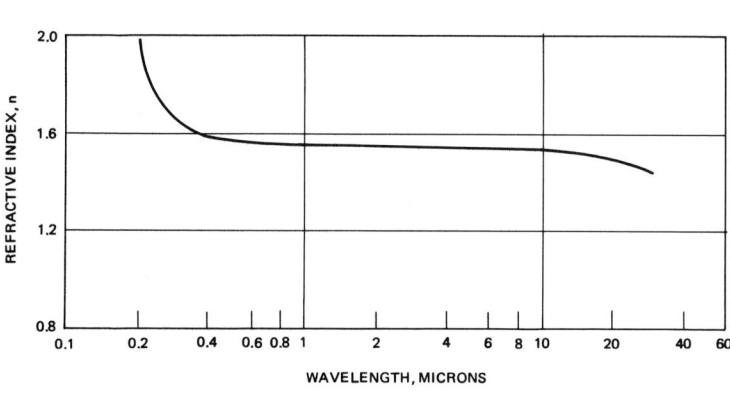

(REF. 3)

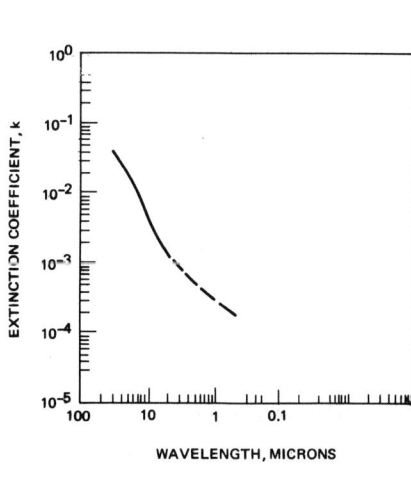

(REF. 4)

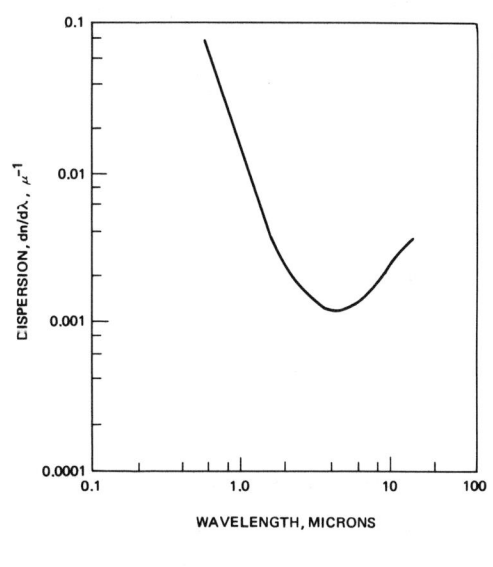

(REF. 5)

REFERENCES:

1. A. Smakula, et al., "Harshaw Optical Crystals", The Harshaw Chemical Co., Cleveland, (1967).

2. C. H. Perry, et al., M.I.T. Cambridge Res. Lab. Electronics, Report No. QPR-91, (1968).

3. A. Smakula, Opt. Acta, 9, 205-222, (1962).

4. J. C. Owens, Phys. Rev., 181, 1228-1236, (1969).

5. W. L. Wolfe, Ed. "Handbook of Military Infrared Technology", U.S. Government Printing Office, Washington, D.C., (1965).

POTASSIUM CHLORIDE

OPTICAL MATERIALS PROPERTIES
DATA SHEET

MATERIAL: POTASSIUM CHLORIDE

INTRODUCTION: This data sheet summarizes properties of single crystal potassium chloride.

PHYSICAL PROPERTIES, (298°K)

Density, (g/cm³) 1.99

Melting/Softening Temp. (°K) 1049

Solubility in Water, (g./100 g. H_2O) 34.35 (293°K)

MECHANICAL PROPERTIES, (298°K)

Young's Modulus, (psi) 4.30×10^{-6}

Hardness, (Knoop) 9.3 <100> (200g)

THERMAL PROPERTIES, (298°K)

Linear Expansion Coeff. (°K⁻¹) 36×10^{-6}

Thermal Conductivity (10^{-4} cal/(cm sec °K)) 156 (315°K)

Specific Heat, (cal/g)/°K 0.162 (273°K)

OPTICAL PROPERTIES, (298°K)

Dispersion Equation:
(for the ultraviolet and visible)

$$n^2 = a^2 + \frac{M_1}{\lambda^2 - \lambda_1^2} + \frac{M_2}{\lambda^2 - \lambda_2^2} - k\lambda^2 - h\lambda^4, \quad n^2 = b^2$$

$$+ \frac{M_1}{\lambda^2 - \lambda_1^2} - \frac{M_2}{\lambda^2 - \lambda_2^2} + \frac{M_3}{\lambda_3^2 - \lambda^2}$$

$a^2 = 2.174967$ $k = 0.000513495$

$M_1 = 0.008344206$ $h = 0.06167587$

$\lambda_1^2 = 0.0119082$ $b^2 = 3.866619$

$M_2 = 0.00698382$ $M_3 = 5569.715$

$\lambda_2^2 = 0.0255550$ $\lambda_3^2 = 3292.47$

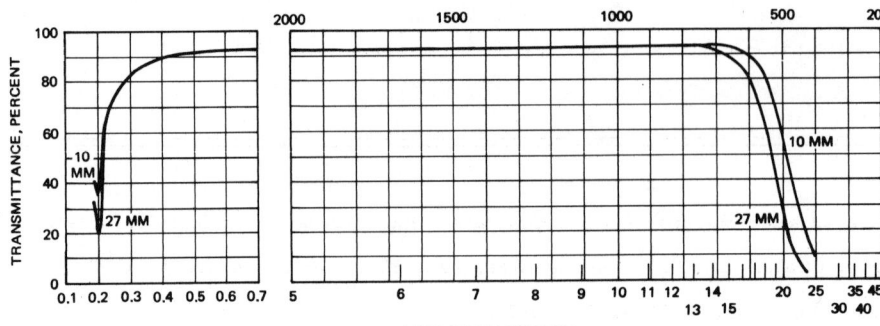

(REF. 1)

MATERIAL: POTASSIUM CHLORIDE

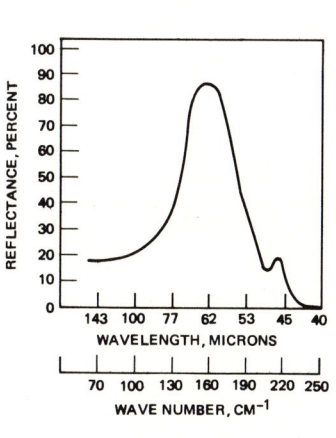

(REF. 2)

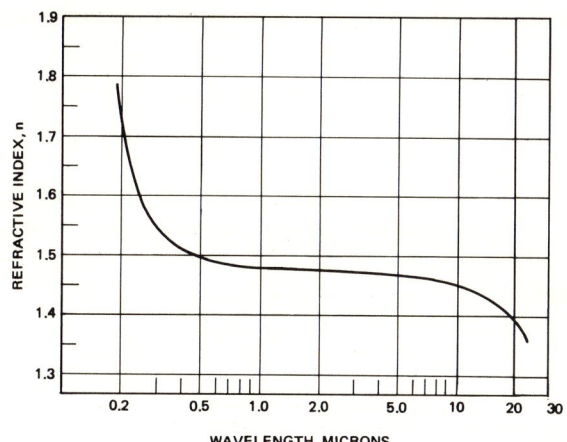

(REF. 3)

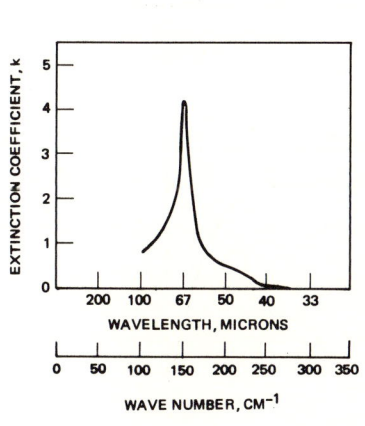

(REF. 4)

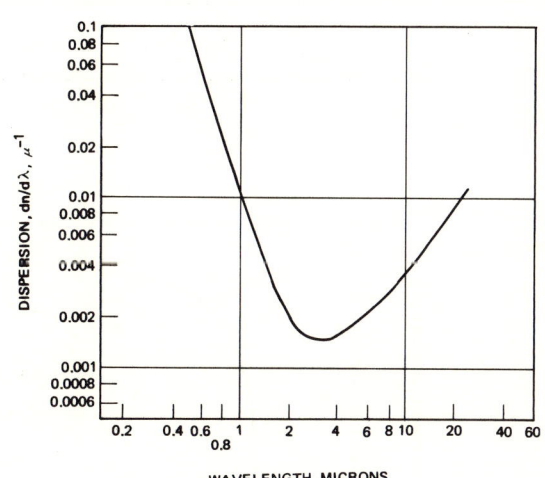

(REF. 5)

REFERENCES:

1. A. Smakula, et al., "Harshaw Optical Crystals", The Harshaw Chemical Co., Cleveland, (1967).

2. C. H. Perry, et al., MIT - Cambridge Res. Lab. Electronics, Report No. QPR-91, (1968).

3. J. A. Mauro, "Optical Engineering Handbook", General Electric Co., Scranton, Pa., (1963).

4. J. N. Plendl and P. J. Gielisse, Appl. Optics, 3, 943-949, (1964).

5. A. Smakula, Opt. Acta, 9, 205-222, (1962).

POTASSIUM IODIDE

OPTICAL MATERIALS PROPERTIES
DATA SHEET

MATERIAL: POTASSIUM IODIDE

INTRODUCTION: This data sheet presents properties of single crystal potassium iodide.

PHYSICAL PROPERTIES, (298°K)

Density, (g/cm^3)	3.13
Melting/Softening Temp. (°K)	996
Solubility in Water, (g./100 g. H$_2$O)	144.5 (293°K)

MECHANICAL PROPERTIES, (298°K)

Young's Modulus, (psi)	4.57 x 10^{-6}
Hardness, (Knoop)	5

THERMAL PROPERTIES, (298°K)

Linear Expansion Coeff. (°K^{-1})	42.6 x 10^{-6} (313°K)
Thermal Conductivity (10^{-4} cal/(cm sec °K)	50.1
Specific Heat, (cal/g)/°K	0.075 (270°K)

OPTICAL PROPERTIES, (298°K)

Dispersion Equation:	Not available
Transmission Region, (External Transmittance ≥10% with 2.0 mm. thickness)	0.25 - 45μ

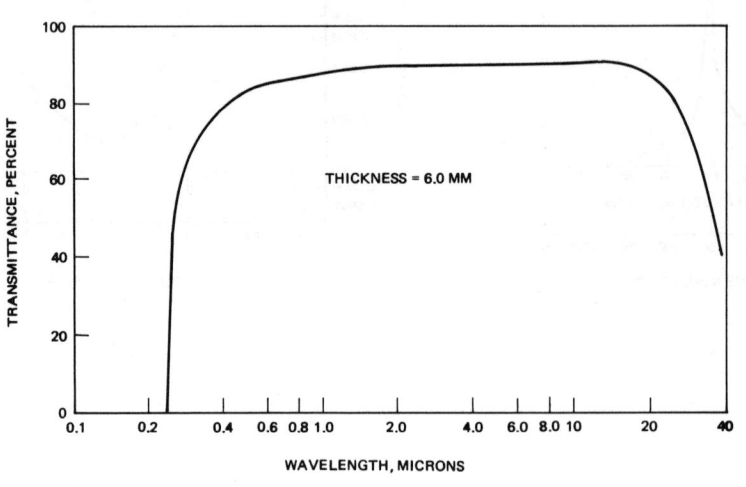

(REF. 1)

MATERIAL: POTASSIUM IODIDE

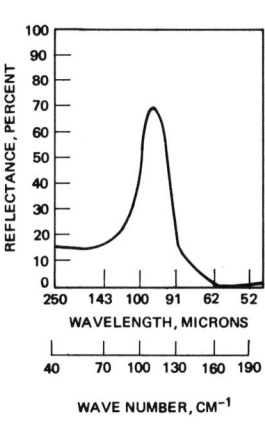

(REF. 2)

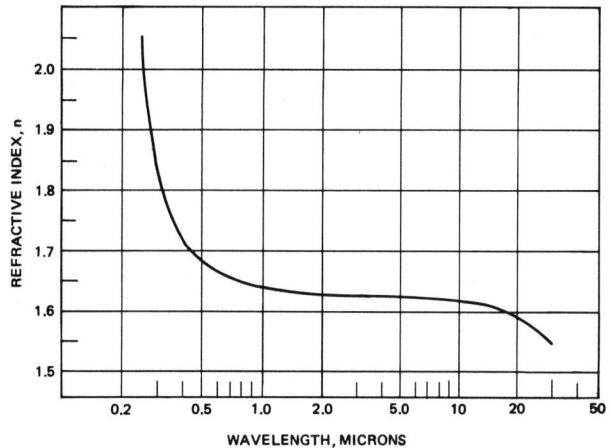

(REF. 3)

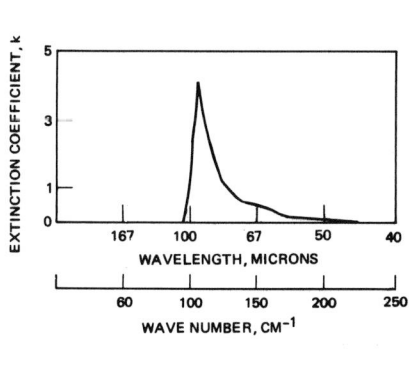

(REF. 4)

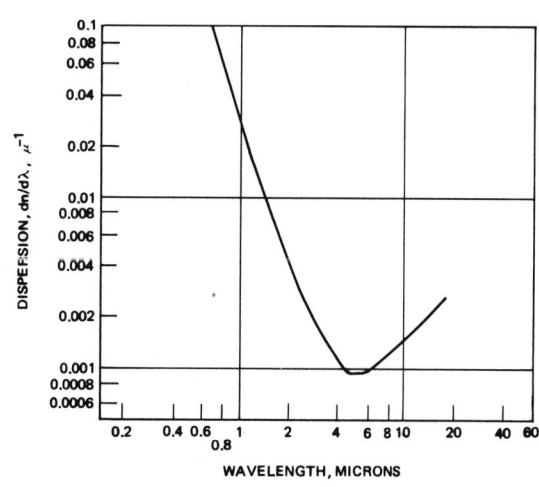

(REF. 5)

REFERENCES:

1. Isomet Corp., "Optical Crystals", Bull. No. 1101, (1965).

2. C. H. Perry, et al., MIT, Cambridge Res. Lab. of Electronics, Report No. QPR-91, (1968).

3. J. A. Mauro, "Optical Engineering Handbook", General Electric Co., Scranton, Pa., (1963).

4. A. Hadni, et al., Appl. Optics, 7, 161-165, (1968).

5. A. Smakula, Opt. Acta, 9, 205-222, (1962).

SELENIUM
(Amorphous)

OPTICAL MATERIALS PROPERTIES MATERIAL: SELENIUM
DATA SHEET   (Amorphous Film)

INTRODUCTION: This data sheet presents properties of amorphous, vacuum-deposited selenium film.

PHYSICAL PROPERTIES, (298°K)

    Density, (g/cm^3)   4.25
    Softening Temp. (°K)   313 - 323
    Solubility in Water,
    (g./100 g. H_2O)   <0.005

MECHANICAL PROPERTIES, (298°K)

    Young's Modulus, (psi)   not available
    Hardness, (Knoop)   not available

THERMAL PROPERTIES, (298°K)

    Linear Expansion
    Coeff. (°K)$^{-1}$ 37 x 10^{-6}
    Thermal Conductivity
    (10^{-4} cal/(cm sec °K)) 31
    Specific Heat,
    (cal/g)/°K   0.08

OPTICAL PROPERTIES, (298°K)

    Dispersion Equation: not available
    Transmission Region,
    (External Transmittance
    ≥10% with 2.0 mm.
    thickness)   1.0 - 20μ

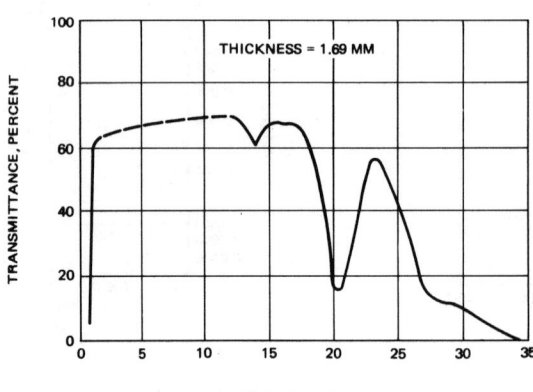

(REF. 1)

MATERIAL: SELENIUM
(Amorphous Film)

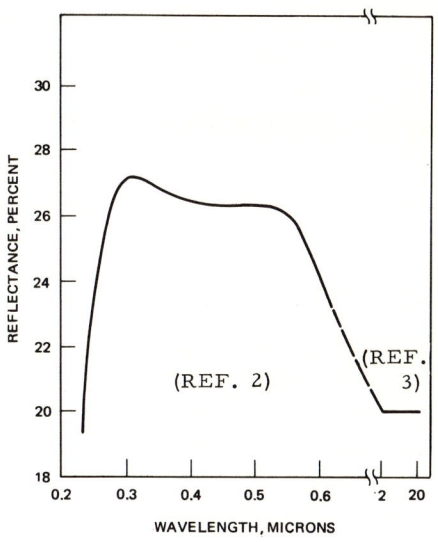

(REF. 2) (REF. 3)

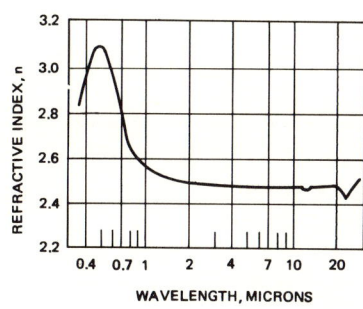

(REF. 4)

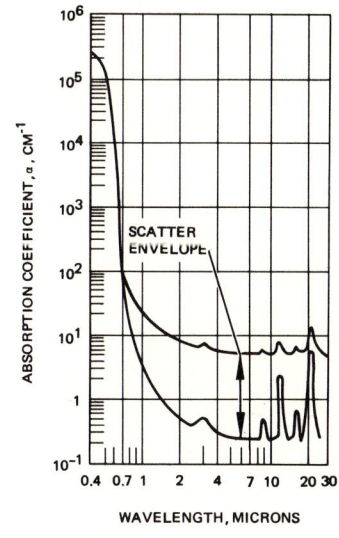

(REF. 4)

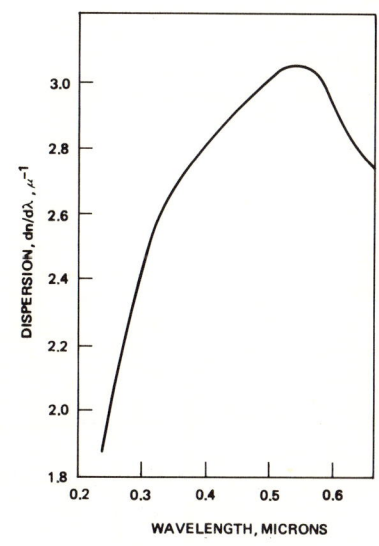

(REF. 2)

REFERENCES:

1. R.S. Caldwell, Special Report on Contract DA 36-039-SC-71131, Purdue University, (1958).

2. W.F. Koehler, et al., J. Opt. Soc. Am., 49, 109-115, (1959).

3. A. Vasko, Czech. J. Phys., 15, 170-177, (1965).

4. H. Gobrecht and A. Tausend, Z. Physik, 161, 205-220, (1961).

SELENIUM
(Hexagonal)

OPTICAL MATERIALS PROPERTIES DATA SHEET   MATERIAL: <u>SELENIUM - HEXAGONAL</u>

INTRODUCTION: This data sheet summarizes properties of single crystal hexagonal selenium.

PHYSICAL PROPERTIES, (298°K)
- Density, (g/cm^3)   4.82
- Melting/Softening Temp. (°K)   490
- Solubility in Water, (g./100 g. H$_2$O)   <0.005

MECHANICAL PROPERTIES, (298°K)
- Young's Modulus, (psi) 8.4×10^6
- Hardness, (Mohs)   2.0

THERMAL PROPERTIES, (298°K)
- Linear Expansion Coeff., (°K)$^{-1}$ 37.9
- Thermal Conductivity (10^{-4} cal/(cm sec °K)   60
- Specific Heat, (cal/g)/°K   0.076

OPTICAL PROPERTIES, (298°K)
- Dispersion Equation   Not available
- Transmission Region, (External Transmittance ≥10% with _____ mm. thickness)   Not available

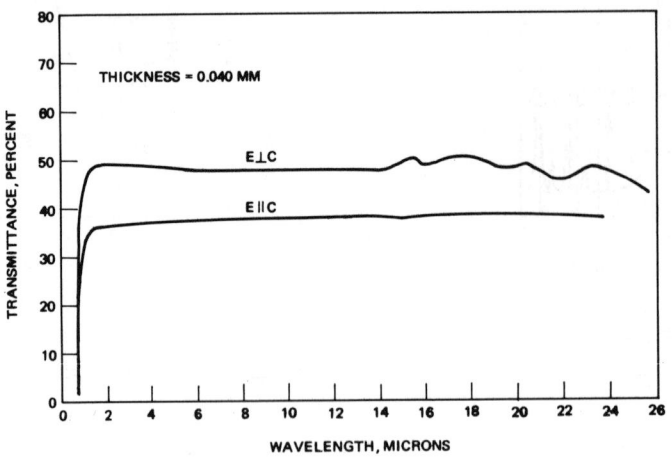

(REF. 1)

MATERIAL: SELENIUM - HEXAGONAL

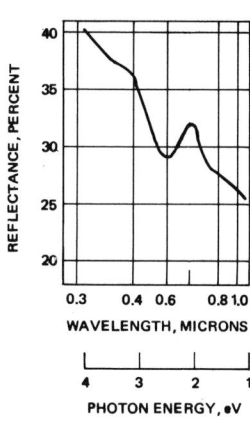

(REF. 2)

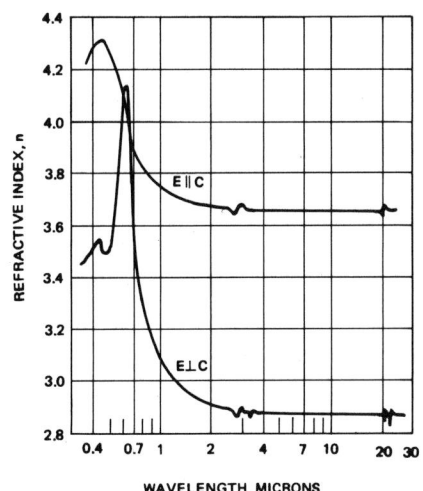

(REF. 3)

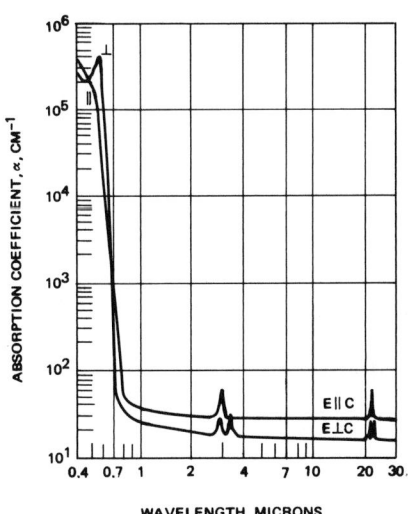

(REF. 3)

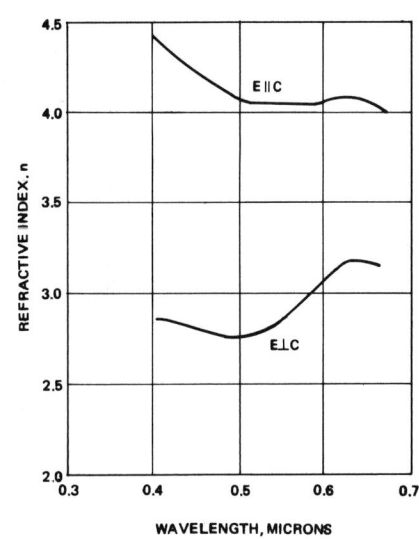

(REF. 4)

REFERENCES:

1. Purdue University, U.S. Govt. Report No. AD 54914, (1954).

2. J. Stuke, Z. Physik, 134, 194-207, (1953).

3. H. Gobrecht and A. Tausend, Z. Physik, 161, 205-220, (1961).

4. C.H. Skinner, Phys. Rev., 9, 148-159, (1917).

SILICA
(Crystalline)

OPTICAL MATERIALS PROPERTIES MATERIAL: SILICA-
DATA SHEET   CRYSTALLINE

INTRODUCTION: This data sheet contains information for high purity crystalline silica.

PHYSICAL PROPERTIES, (298°K)
- Density, (g/cm^3) _____ 2.65
- Melting/Softening Temp. (°K) 1743
- Solubility in Water, (g./100 g. H$_2$0) _____ < 0.001

MECHANICAL PROPERTIES, (298°K)*
- Young's Modulus, (psi) (14.1-11.1) x 10
- Hardness, (Knoop) _____ 741 (500 g.)

THERMAL PROPERTIES, (298°K)*
- Linear Expansion Coeff. (°K^{-1}) (7.97-13.37) x 10^{-6}
- Thermal Conductivity (10^{-4} cal/(cm sec °K)) 255 - 148
- Specific Heat, (cal/g)/°K _____ 0.188

*Double values for E ∥ C and E ⊥ C, respectively.

OPTICAL PROPERTIES, (298°K)
- Dispersion Equation: not available
- Transmission Region, (External Transmittance ≥10% with 2.0 mm. thickness) 0.12 - 4.5μ

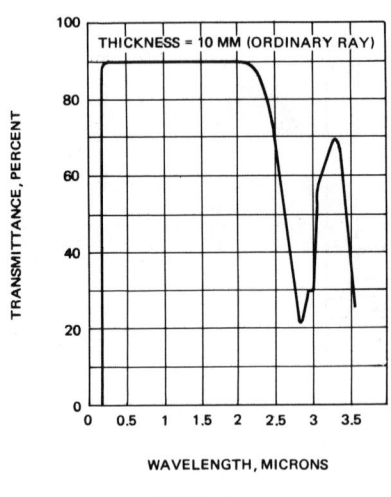

(REF. 1)

MATERIAL: <u>SILICA-
CRYSTALLINE</u>

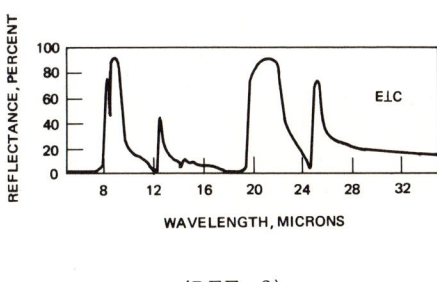

(REF. 2)

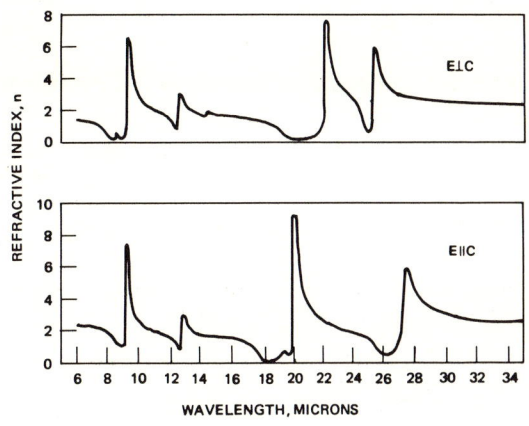

(REF. 2)

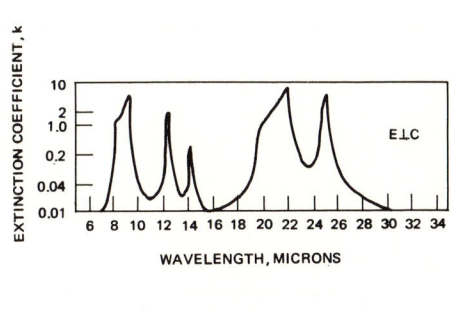

(REF. 2)

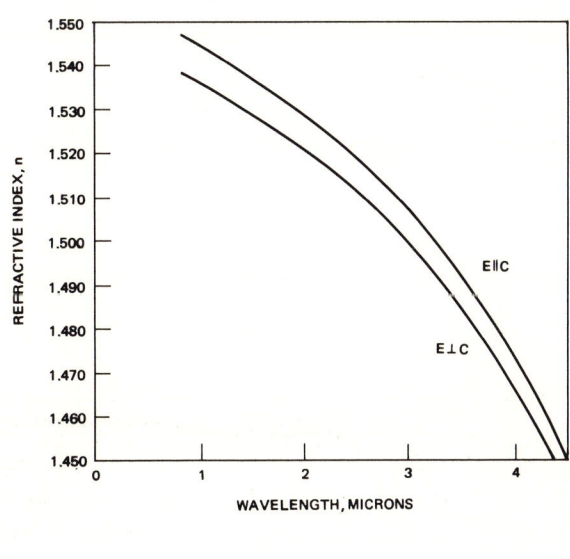

(REF. 3)

REFERENCES:

1. P. Billard, Acta Electronica, <u>6</u>, 75 - 169, (1962).

2. W. G. Spitzer and D. A. Kleinman, Phys. Rev., <u>121</u>, 1324 - 1335, (1961).

3. American Institute of Physics Handbook, McGraw-Hill Book Co., Inc., New York, (1957).

SILICA
(Fused)

OPTICAL MATERIALS PROPERTIES DATA SHEET   MATERIAL: SILICA-FUSED

INTRODUCTION: This data sheet summarizes properties of high purity fused silica.

PHYSICAL PROPERTIES, (298°K)

Density, (g/cm^3)   2.20

Melting/Softening Temp. (°K)   1943

Solubility in Water, (g./100 g. H$_2$0)   <0.001

MECHANICAL PROPERTIES, (298°K)

Young's Modulus, (psi)   10.6 x 10^6

Hardness, (Knoop)   461 (200 g.)

THERMAL PROPERTIES, (298°K)

Linear Expansion Coeff. (°K^{-1})   0.55 x 10^{-6}

Thermal Conductivity (10^{-4}cal/(cm sec °K)   33.0

Specific Heat, (cal/g)/°K   0.18

OPTICAL PROPERTIES, (298°K)

Dispersion Equation:

$$n^2 = 2.978645 + \frac{0.008777808}{\lambda^2 - 0.010609} - \frac{84.06224}{96.00000 - \lambda^2}$$

Transmission Region, (External Transmittance ≥10% with 2.0 mm. thickness)   0.12 - 4.5μ

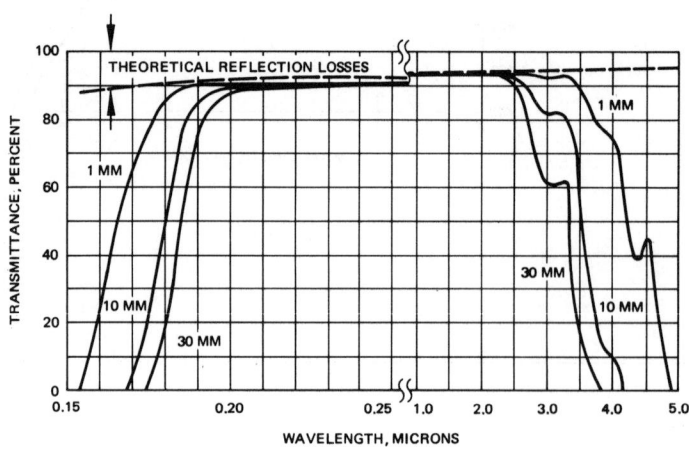

(REF. 1)

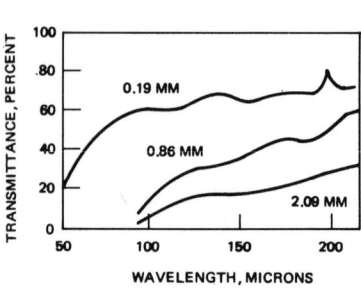

(REF. 2)

MATERIAL: SILICA-FUSED

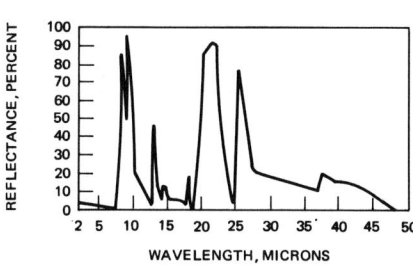

(REF. 3)

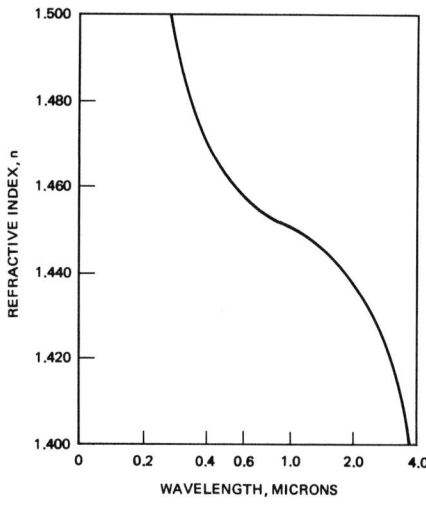

(REF. 4)

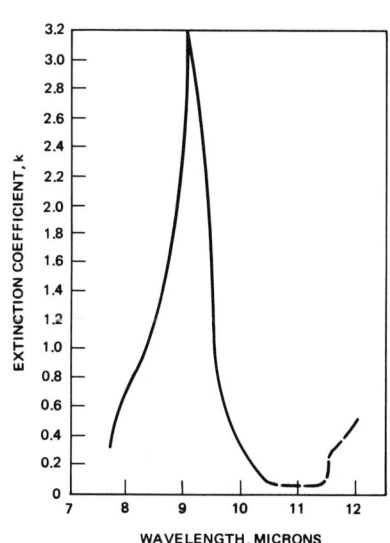

(REF. 5)

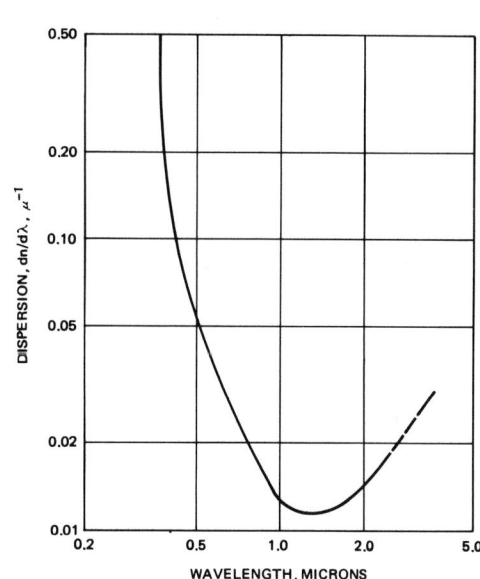

(REF. 6)

REFERENCES:

1. Amersil, Inc. Data Sheet.
2. C. H. Cartwright, Z. Physik, 90, 480-488, (1934).
3. D. E. McCarthy, Appl. Optics, 2, 591-595, (1963).
4. I. H. Malitson, J. Opt. Soc. Am., 55, 1205-1209, (1965).
5. G. W. Cleek, Appl. Optics, 5, 771-776, (1966).
6. W. S. Rodney and R. J. Spindler, J. Res. NBS, 53, 185-189, (1964).

SILICON

OPTICAL MATERIALS PROPERTIES DATA SHEET   MATERIAL: SILICON

INTRODUCTION: This data sheet covers properties of single crystal silicon.

PHYSICAL PROPERTIES, (298°K)

Density, (g/cm³)   2.33

Melting/Softening Temp. (°K)   1693

Solubility in Water, (g./100 g. H_2O)   < 0.005

MECHANICAL PROPERTIES, (298°K)

Young's Modulus, (psi) 19.0×10^6

Hardness, (Knoop)   1100 - 1400

THERMAL PROPERTIES, (298°K)

Linear Expansion Coeff. (°K⁻¹) 4.7×10^{-6}

Thermal Conductivity (10^{-4} cal/(cm sec °K))   3900 (313°K)

Specific Heat, (cal/g)/°K   0.168

OPTICAL PROPERTIES, (298°K)

Dispersion Equation:

$$n = 3.41696 + 0.138497\,L + 0.013924\,L^2 - 0.0000209\,\lambda^2 + 0.000000148\,\lambda^4$$

where: $L = (\lambda^2 - 0.028)^{-1}$

Transmission Region, (External Transmittance ≥10% with 2.0 mm. thickness)   1.2 - 15µ

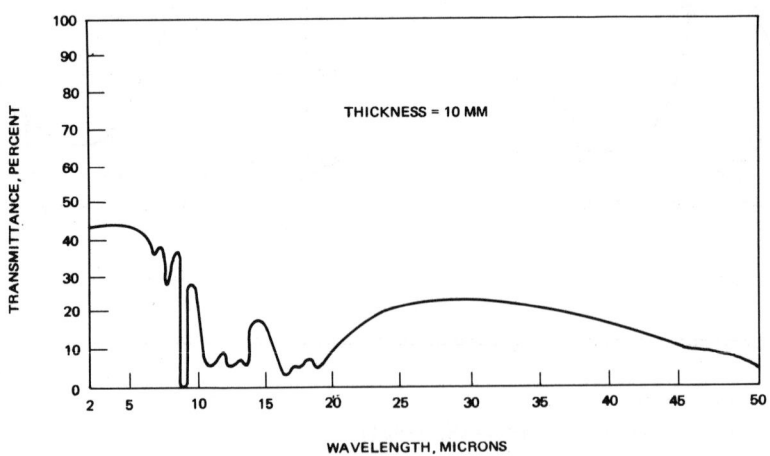

(REF. 1)

MATERIAL: SILICON

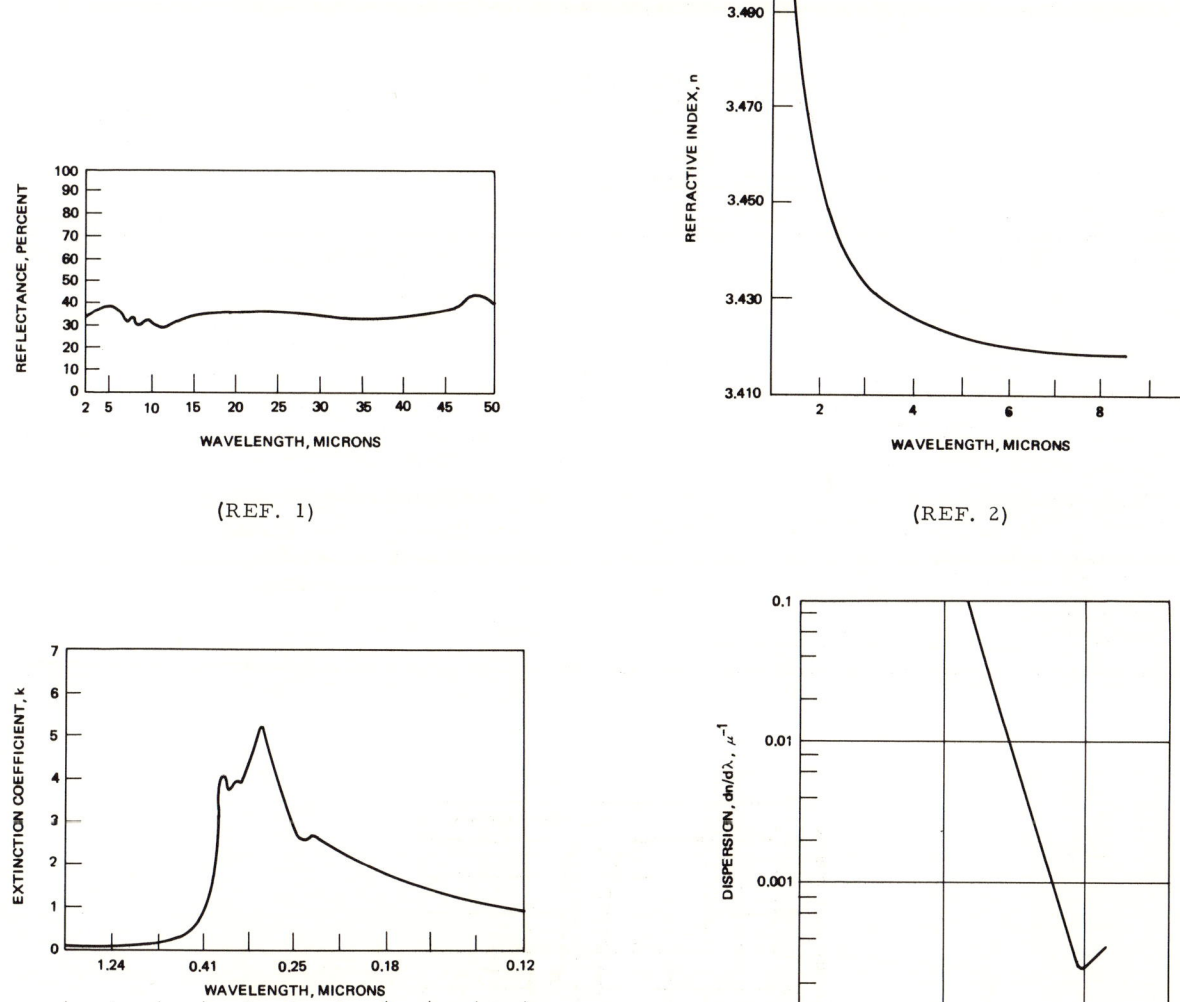

(REF. 1)

(REF. 2)

(REF. 3)

(REF. 4)

REFERENCES:

1. D. E. McCarthy, Appl. Optics, 2, 591-603, (1963).

2. C. D. Salzberg and J. J. Villa, J. Opt. Soc. Am., 47, 244-246, (1957).

3. H. W. Verleur, J. Opt. Soc. Am., 58, 1356-1364, (1968).

4. S. S. Ballard, et al., "Optical Materials for Infrared Instrumentation", Report No. AD217367, (1959).

SILICON CARBIDE

OPTICAL MATERIALS PROPERTIES DATA SHEET   MATERIAL: SILICON CARBIDE (Type 6H)

INTRODUCTION: This data sheet contains information for single hexagonal silicon carbide crystals, type 6H.

PHYSICAL PROPERTIES, (298°K)
- Density, (g/cm³)   3.21
- Melting/Softening Temp. (°K)   3103
- Solubility in Water, (g./100 g. H_2O)   <0.01

MECHANICAL PROPERTIES, (298°K)
- Young's Modulus, (psi) 56×10^6
- Hardness, (Knoop)   2130 - 2755

THERMAL PROPERTIES, (298°K)
- Linear Expansion Coeff., (°K)⁻¹ $\sim 4 \times 10^{-6}$
- Thermal Conductivity (10^{-4} cal/cm sec °K)   1000
- Specific Heat, (cal/g)/°K   0.165

OPTICAL PROPERTIES, (298°K)
- Dispersion Equation   Not available
- Transmission Region, (External Transmittance ≥10% with _____ mm. thickness)   Not available

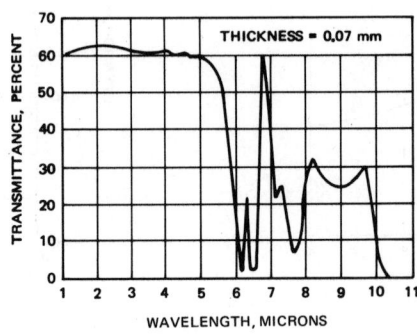

(REF. 1)

MATERIAL: <u>SILICON CARBIDE</u>
(Type 6H)

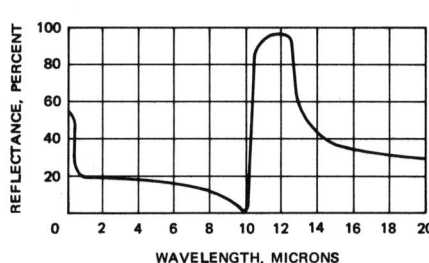

(REF. 2, 3)

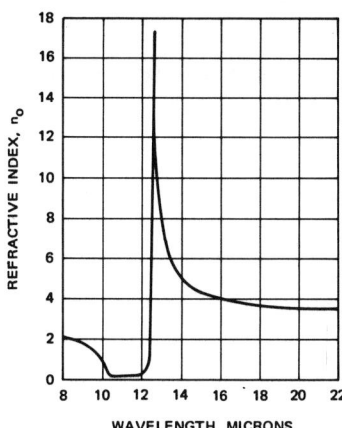

(REF. 2)

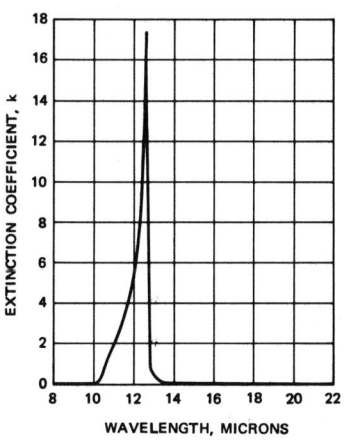

(REF. 2)

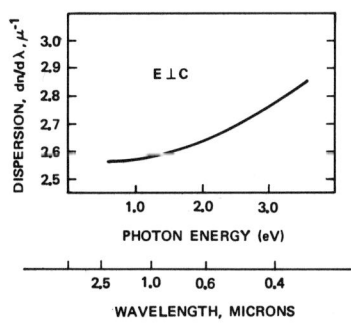

(REF. 4)

REFERENCES:

1. H.G. Lipson, Conf. on Silicon Carbide, Boston, 1959, Silicon Carbide - A High Temperature Semiconductor; Proc., Ed. J. R. O'Connor and J. Smittens, Pergamon Press, N.Y., (1960).

2. W. G. Spitzer, et al, Phys. Rev., <u>113</u>, 127-132, (1959).

3. H.R. Philipp and E.A. Taft, Conf. on Silicon Carbide, Boston, 1959, Silicon Carbide - A High Temperature Semiconductor, Proc. Ed., J. R. O'Connor and J. Smittens, Pergamon Press, N.Y., (1960).

4. W.J. Choyke and L. Patrick, J. Opt. Soc. Am., <u>58</u>, 377-379, (1968).

SILVER

OPTICAL MATERIALS PROPERTIES DATA SHEET   MATERIAL: SILVER (Film)

INTRODUCTION: This data sheet presents properties of vacuum-evaporated silver film, generally on a glass substrate.

PHYSICAL PROPERTIES, (298°K)		MECHANICAL PROPERTIES, (298°K)	
Density, (g/cm^3)	10.5	Young's Modulus, (psi)	8.27×10^5
Melting/Softening Temp. (°K)	1234	Hardness, (Rockwell F)	91 (bulk)
Solubility in Water, (g./100 g. H$_2$O)	<0.001		

THERMAL PROPERTIES, (298°K)		OPTICAL PROPERTIES, (298°K)	
Linear Expansion Coeff., (°K)$^{-1}$	19.1×10^{-6}	Dispersion Equation	Not available
Thermal Conductivity 10^{-4} cal/(cm sec °K)	1.0×10^4	Transmission Region, (External Transmittance ≥10% with ___ mm. thickness)	Not available
Specific Heat, (cal/g)/°K	0.056		

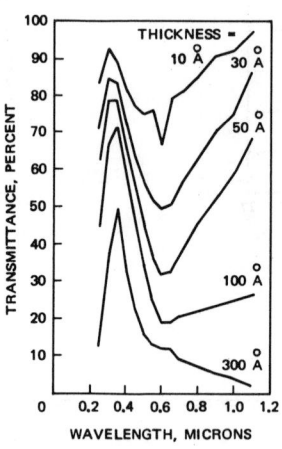

(REF. 1)

MATERIAL: SILVER (Film)

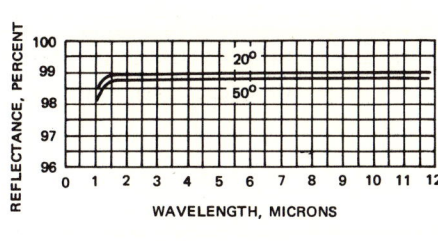

(REF. 2)

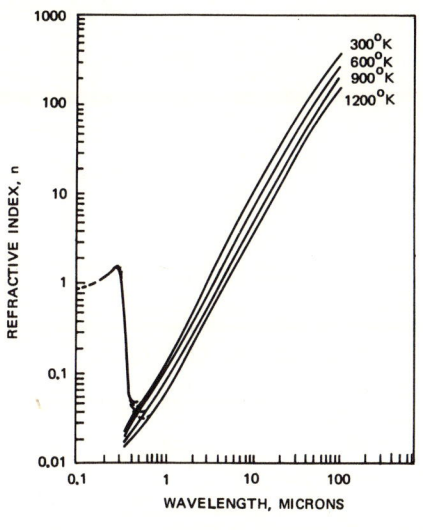

(REF. 3)

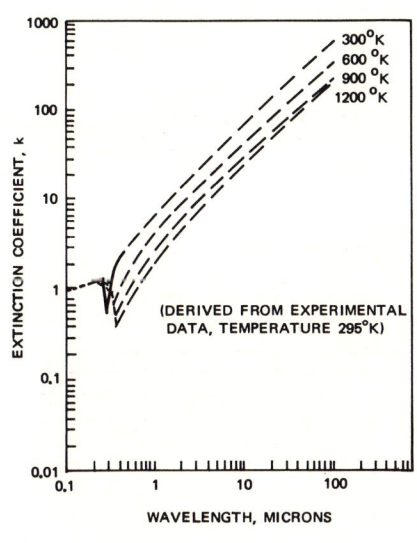

(REF. 3)

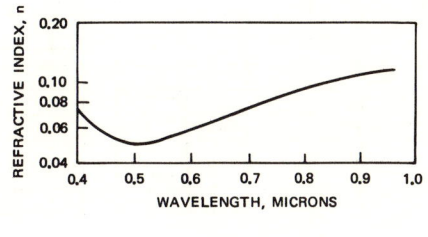

(REF. 4)

REFERENCES:

1. F. Goos, Z. Phys., 100, 95-112, (1936).

2. D.M. Gates, et al, J. Opt. Soc. Am., 48, 88-89, (1958).

3. A.F. Grenis, U.S. Government Report No. AMRA TR 67-02, (1967).

4. L.G. Schulz and F.R. Tangherlini, J. Opt. Soc. Am., 44, 362-368, (1965).

SILVER BROMIDE

OPTICAL MATERIALS PROPERTIES DATA SHEET   MATERIAL: SILVER BROMIDE

INTRODUCTION: This data sheet contains information for single crystal silver bromide.

PHYSICAL PROPERTIES, (298°K)
 Density, (g/cm^3) 6.47
 Melting/Softening Temp. (°K) 705
 Solubility in Water, (g./100 g. H$_2$O) 12.6 x 10^{-6} (293°K)

MECHANICAL PROPERTIES, (298°K)
 Young's Modulus, (psi) 4.64 x 10^6
 Hardness, (Knoop) Not available

THERMAL PROPERTIES, (298°K)
 Linear Expansion Coeff., (°K)$^{-1}$ 34.8 x 10^{-6}
 Thermal Conductivity (10^{-4} cal/(cm sec °K) 29 (273°K)
 Specific Heat, (cal/g)/°K 0.070

OPTICAL PROPERTIES, (298°K)
 Dispersion Equation between 0.54 and 0.65µ:

$$\frac{n^2-1}{n^2+2} = 0.48484 + \frac{0.10279\lambda^2}{\lambda^2 - 0.090000} - 0.004796\lambda^2$$

 Transmission Region, (External Transmittance ≥10% with 2.0 mm. thickness) 0.45 - 35µ

(REF. 1)

MATERIAL: SILVER BROMIDE

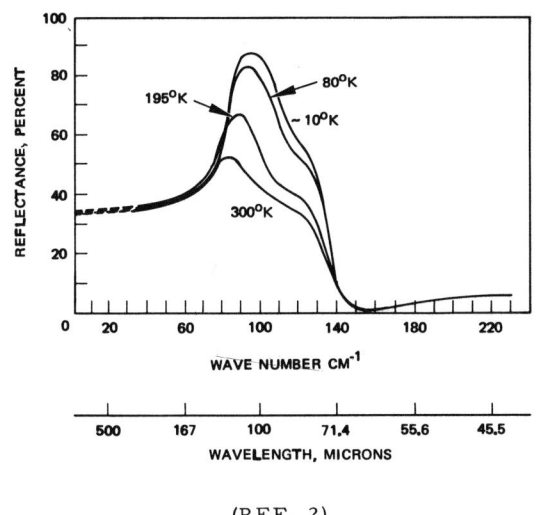

(REF. 2)

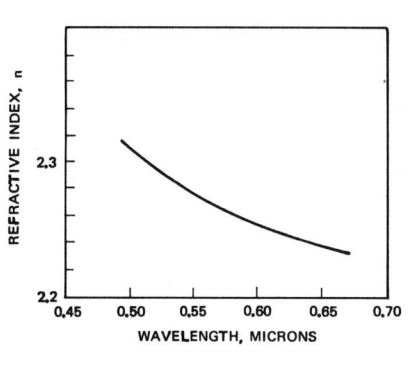

(REF. 3)

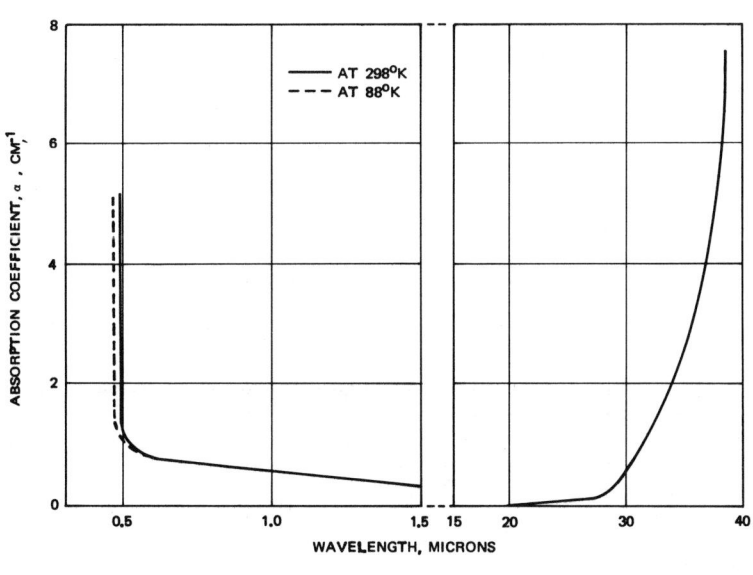

(REF. 4)

REFERENCES:

1. A. Smakula, et al, "Harshaw Optical Crystals," Harshaw Chemical Co., Cleveland, (1967).

2. C.H. Perry, Report No. AD 679 213, (1968).

3. H. Schroeder, Z. Physik, 67, 24-36, (1931).

4. A. Smakula, Report No. AD 663 734, (1967).

SILVER CHLORIDE

OPTICAL MATERIALS PROPERTIES DATA SHEET MATERIAL: SILVER CHLORIDE

INTRODUCTION: This data sheet summarizes properties of single crystal silver chloride.

PHYSICAL PROPERTIES, (298°K)

Density, (g/cm^3)   5.56

Melting/Softening Temp. (°K)   731

Solubility in Water, (g./100 g. H_2O) 1.5×10^{-4}

MECHANICAL PROPERTIES, (298°K)

Young's Modulus, (psi) 2.9×10^6

Hardness, (Knoop)   9.5 (200 g)

THERMAL PROPERTIES, (298°K)

Linear Expansion Coeff., (°K)$^{-1}$ 30×10^{-6} (293°K)

Thermal Conductivity (10^{-4} cal/(cm sec °K))   26 (273°K)

Specific Heat, (cal/g)/°K   0.0848 (273°K)

OPTICAL PROPERTIES, (298°K)

Dispersion Equation

$$n^2 = 4.00804 - 0.00085111\lambda^2 - 0.00000019762\lambda^4 + 0.079086/(\lambda^2 - 0.04584)$$

Transmission Region, (External Transmittance ≥10% with 2.0 mm. thickness)   0.4 - 2.8μ

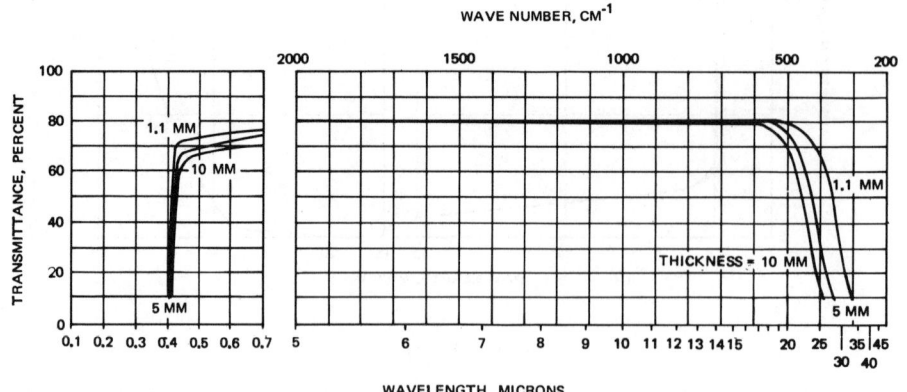

(REF. 1)

MATERIAL: SILVER CHLORIDE

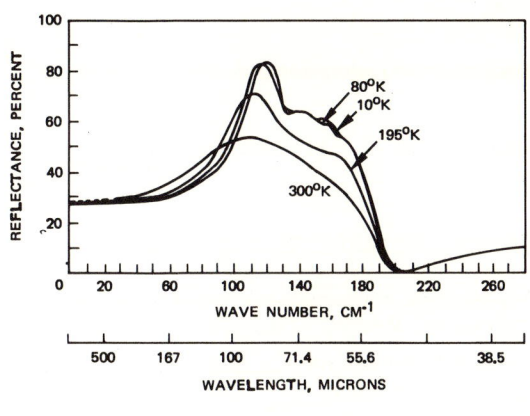

(REF. 2)

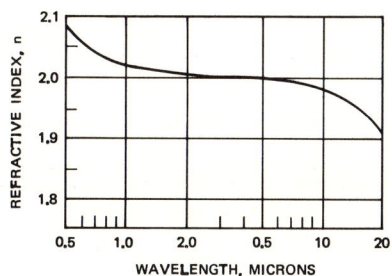

(REF. 3)

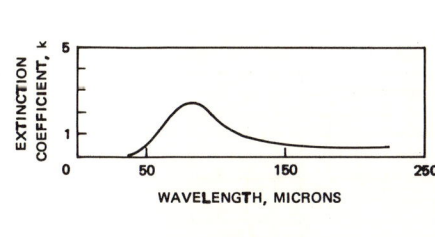

(REF. 4)

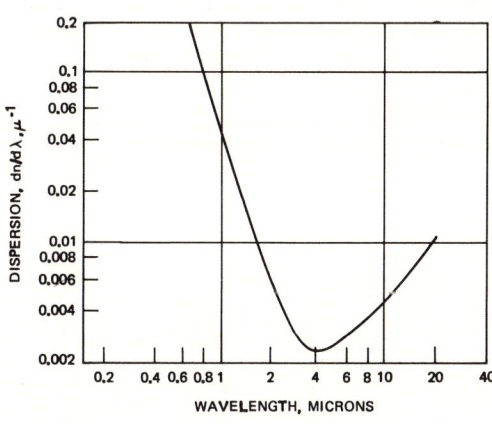

(REF. 5)

REFERENCES:

1. A. Smakula, et al, "Harshaw Optical Crystals," Harshaw Chemical Co., Cleveland, (1967).

2. C.H. Perry, Report No. AD 679 213, (1968).

3. J.A. Mauro, "Optical Engineering Handbook," General Electric Co., Scranton, Pa., (1963).

4. A. Hadni, et al., J. de Phys., 28, C1-118 - C1-128, (1967).

5. A. Smakula, Opt. Acta, 9, 205-222, (1962).

SODIUM CHLORIDE

OPTICAL MATERIALS PROPERTIES DATA SHEET MATERIAL: <u>SODIUM CHLORIDE</u>

INTRODUCTION: <u>This data sheet summarizes properties of single crystal sodium chloride.</u>

PHYSICAL PROPERTIES, (298°K)
- Density, (g/cm^3)   <u>3.16</u>
- Melting/Softening Temp. (°K)   <u>1090</u>
- Solubility in Water, (g./100 g. H_2O)   <u>36.2</u>

MECHANICAL PROPERTIES, (298°K)
- Young's Modulus, (psi)   <u>5.8 x 10^6</u>
- Hardness, (Knoop)   <u>17</u>

THERMAL PROPERTIES, (298°K)
- Linear Expansion Coeff., (°K)$^{-1}$ <u>44 x 10^{-6}</u>
- Thermal Conductivity (10^{-4} cal/(cm sec °K)   <u>155 (289°K)</u>
- Specific Heat, (cal/g)°K   <u>0.204 (273°K)</u>

OPTICAL PROPERTIES, (298°K)
- Dispersion Equation   <u>Not available</u>
- Transmission Region, (External Transmittance ≥10% with <u>2.0</u> mm. thickness)   <u>0.21 - 26µ</u>

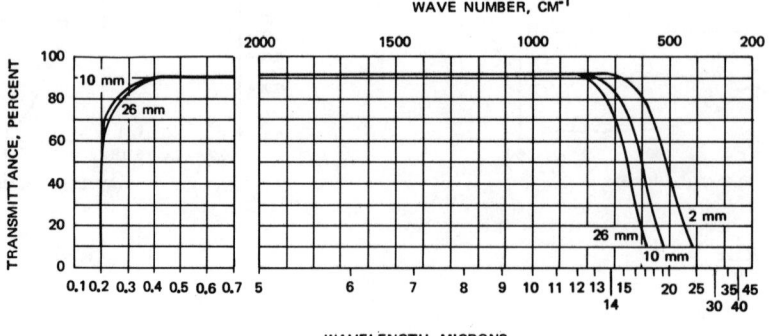

(REF. 1)

MATERIAL: SODIUM CHLORIDE

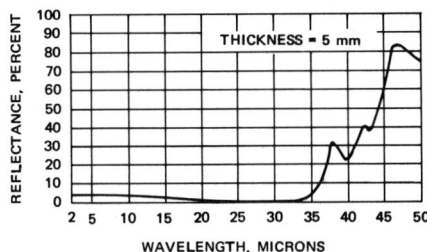

(REF. 2)

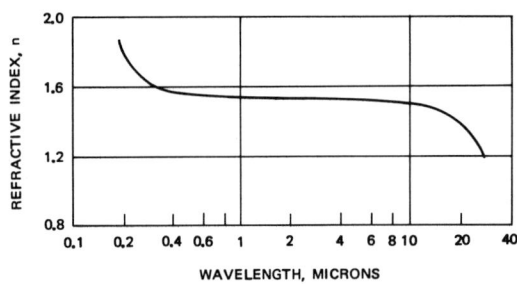

(REF. 3)

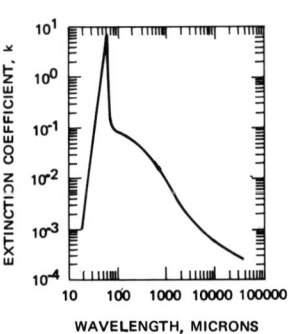

(REF. 4)

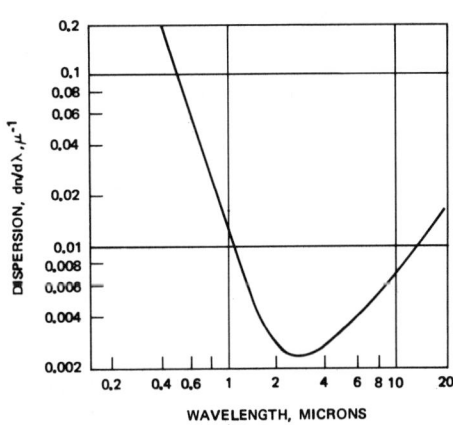

(REF. 3)

REFERENCES:

1. A. Smakula, et al, "Harshaw Optical Crystals," Harshaw Chemical Co., Cleveland, (1967).

2. D.E. McCarthy, Appl. Optics, 2, 591-595, (1963).

3. A. Smakula, Opt. Acta 9, 205-222, (1962).

4. J.C. Owens, Phys. Rev., 181, 1228-1236, (1969).

SODIUM FLUORIDE

OPTICAL MATERIALS PROPERTIES DATA SHEET MATERIAL: SODIUM FLUORIDE

INTRODUCTION: This data sheet summarizes properties of sodium fluoride single crystals.

PHYSICAL PROPERTIES, (298°K)
- Density, (g/cm^3) 2.79
- Melting/Softening Temp. (°K) 1270
- Solubility in Water, (g./100 g. H$_2$O) 4.2 (293°K)

MECHANICAL PROPERTIES, (298°K)
- Young's Modulus, (psi) 1.41×10^7
- Hardness, (Knoop) 60

THERMAL PROPERTIES, (298°K)
- Linear Expansion Coeff., (°K)$^{-1}$ 36×10^{-6}
- Thermal Conductivity (10^{-4} cal/(cm sec °K) 505
- Specific Heat, (cal/g)/°K 0.26 (273°K)

OPTICAL PROPERTIES, (298°K)
- Dispersion Equation Not available
- Transmission Region, (External Transmittance ≥10% with 2.0 mm. thickness) 0.19 - 15µ

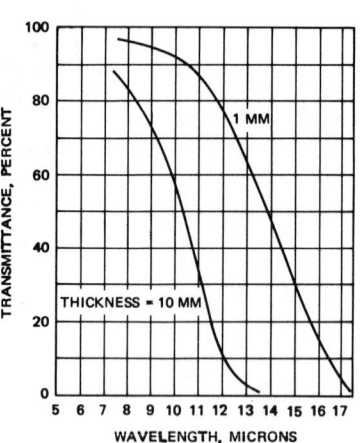

(REF. 1)

MATERIAL: SODIUM FLUORIDE

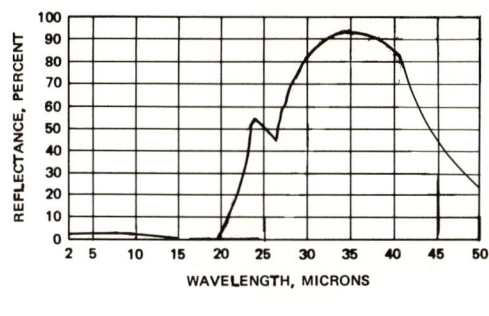

(REF. 2)

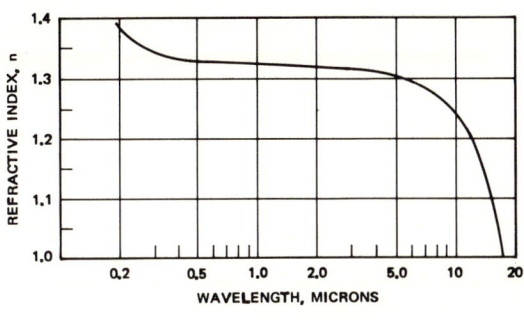

(REF. 3)

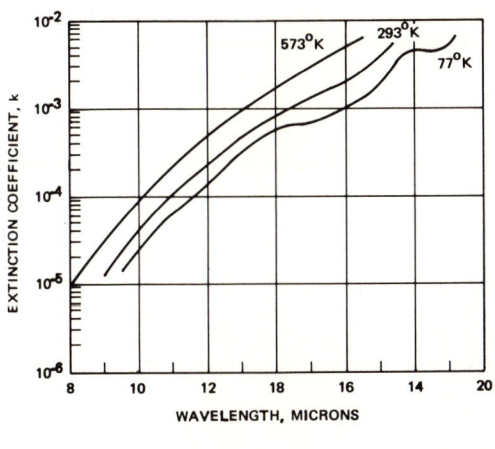

(REF. 4)

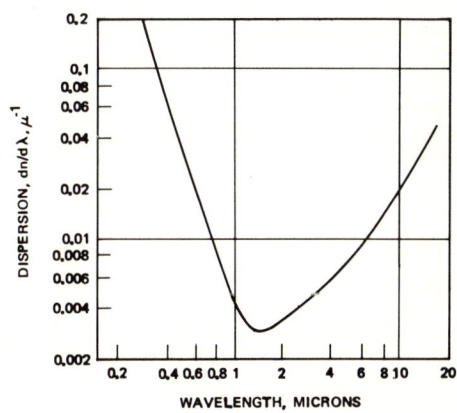

(REF. 5)

REFERENCES:

1. P. Billard, Acta Electronica, 6, 75-169, (1962).

2. D.E. McCarthy, Appl. Optics, 4, 317-320, (1965).

3. J.A. Mauro, "Optical Engineering Handbook," General Electric Co., Scranton, Pa., (1963).

4. M. Klier, Z. Physik, 150, 49-63, (1958).

5. A. Smakula, Opt. Acta, 9, 205-222, (1962).

STRONTIUM TITANATE

OPTICAL MATERIALS PROPERTIES DATA SHEET   MATERIAL: STRONTIUM TITANATE

INTRODUCTION: This data sheet summarizes properties of single crystal strontium titanate.

PHYSICAL PROPERTIES, (298°K)

Density, (g/cm^3)   5.13

Melting/Softening Temp. (°K)   2353

Solubility in Water, (g./100 g. H$_2$O)   <0.01

MECHANICAL PROPERTIES, (298°K)

Young's Modulus, (psi)   Not available

Hardness, (Knoop)   595

THERMAL PROPERTIES, (298°K)

Linear Expansion Coeff., (°K)$^{-1}$ 9.4 x 10^{-6}

Thermal Conductivity (10^{-4} cal/(cm sec °K))   145

Specific Heat, (cal/g)/°K   0.13

OPTICAL PROPERTIES, (298°K)

Dispersion Equation

$$n = A + BL + C\lambda^2 + D\lambda^2 + E\lambda^4$$

between 1.0 and 5.3μ

A	B	C
2.28355	0.035906	+0.001666

D	E
-0.0061335	-0.00001502

Transmission Region, (External Transmittance ≥10% with 2.0 mm. thickness)   0.39 - 6.8μ

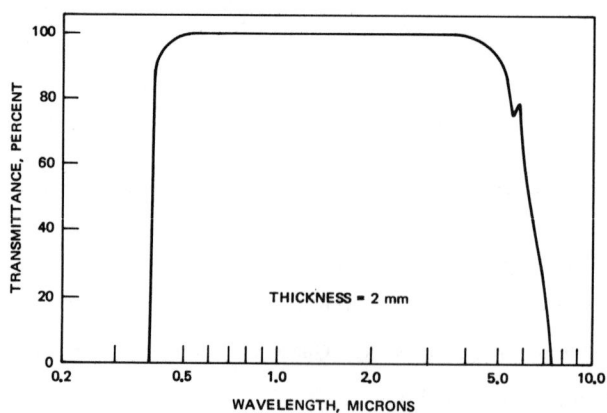

(REF. 1)

MATERIAL: STRONTIUM TITANATE

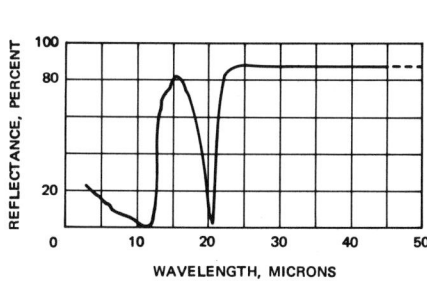

(REF. 2)

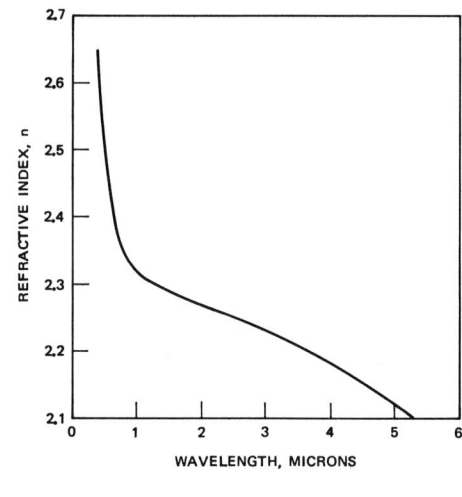

(REF. 3)

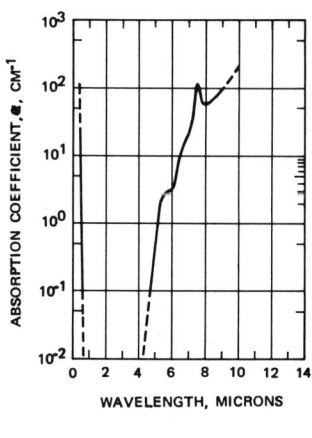

(REF. 4)

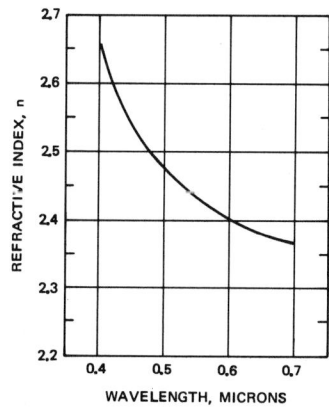

(REF. 4)

REFERENCES:

1. M.D. Beals and L. Merker, Materials in Design Eng., 51, 12-13, (1960).

2. D.E. McCarthy, Appl. Optics, 7, 1997-2000, (1968).

3. S.S. Ballard, et al, Report No. AD 255699, (1961).

4. S.B. Levin, et al, J. Opt. Soc. Am., 45, 737-739, (1955).

TELLURIUM
(Film)

OPTICAL MATERIALS PROPERTIES DATA SHEET   MATERIAL: TELLURIUM (Polycrystalline Film)

INTRODUCTION: This data sheet summarizes properties of polycrystalline tellurium film.

PHYSICAL PROPERTIES, (298°K)
- Density, (g/cm^3)   6.1
- Melting/Softening Temp. (°K)   723
- Solubility in Water, (g./100 g. H$_2$O)   <0.005

MECHANICAL PROPERTIES, (298°K)
- Young's Modulus, (psi)   Not available
- Hardness, (Knoop)   Not available

THERMAL PROPERTIES, (298°K)
- Linear Expansion Coeff., (°K)$^{-1}$ 16.75 x 10^{-6} (313°K)
- Thermal Conductivity (10^{-4} cal/(cm sec °K)   150
- Specific Heat, (cal/g)/°K   0.0479 (573°K)

OPTICAL PROPERTIES (298°K)
- Dispersion Equation   Not available
- Transmission Region, (External Transmittance ≥10% with 2.0 mm. thickness)   3.5 - 8.0µ

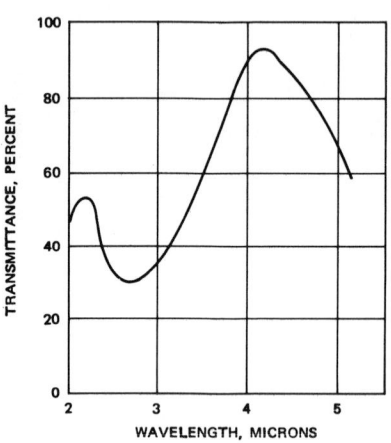

(REF. 1)

MATERIAL: TELLURIUM
(Polycrystalline Film)

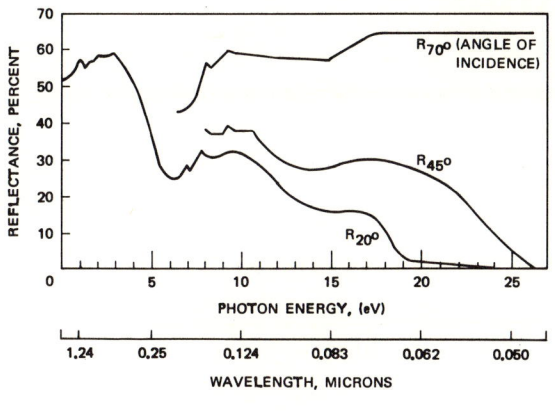
(REF. 2)

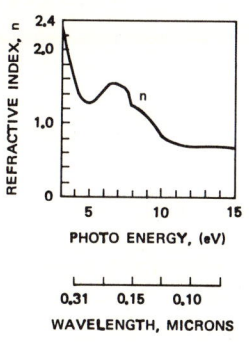
(REF. 2)

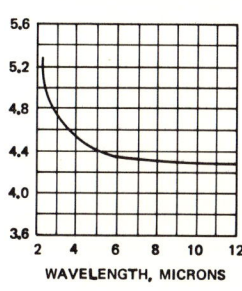
(REF. 1)

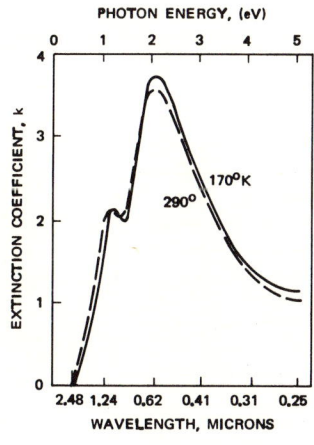
(REF. 3)

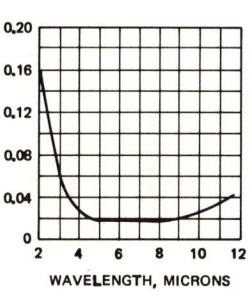
(REF. 1)

REFERENCES:

1. A.S. Valeer and M.A. Gisin, Optics and Spec., 19, 62-65, (1965).

2. J.D. Hayes, et al., J. Appl. Phys, 39, 5527-5532, (1968).

3. H. Keller and J. Stuke, Phys. Stat. Solidi, 8, 831-840, (1965).

TELLURIUM
(Single Crystal)

OPTICAL MATERIALS PROPERTIES DATA SHEET   MATERIAL: TELLURIUM (Single Crystal)

INTRODUCTION: This data sheet presents information information for single crystal tellurium.

PHYSICAL PROPERTIES, (298°K)
- Density, (g/cm^3)   6.25
- Melting/Softening Temp. (°K)   725
- Solubility in Water, (g./100 g. H_2O)   <0.005

MECHANICAL PROPERTIES, (298°K)
- Young's Modulus, (psi)   Not available
- Hardness, (Knoop)   18.4

THERMAL PROPERTIES*, (298°K)
- Linear Expansion Coeff., (°K)$^{-1}$ $(-1.6, +27) \times 10^{-6}$
- Thermal Conductivity (10^{-4} cal/(cm sec °K)   150 (320°K)
- Specific Heat, (cal/g)/°K   0.048

OPTICAL PROPERTIES, (298°K)
- Dispersion Equation   Not available
- Transmission Region, (External Transmittance ≥10% with 2.0 mm. thickness)   3.5 - 8.0µ

*E ∥ C and E ⊥ C axis respectively

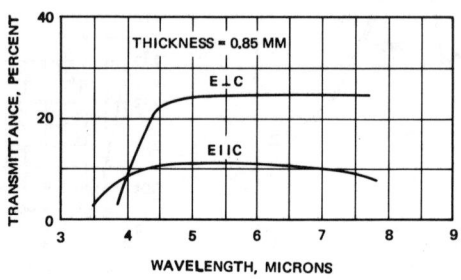

(REF. 1)

MATERIAL: **TELLURIUM**
(Single Crystal)

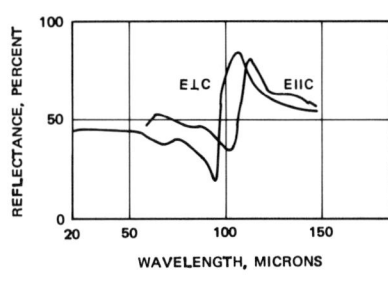

(REF. 2)

(REF. 3)

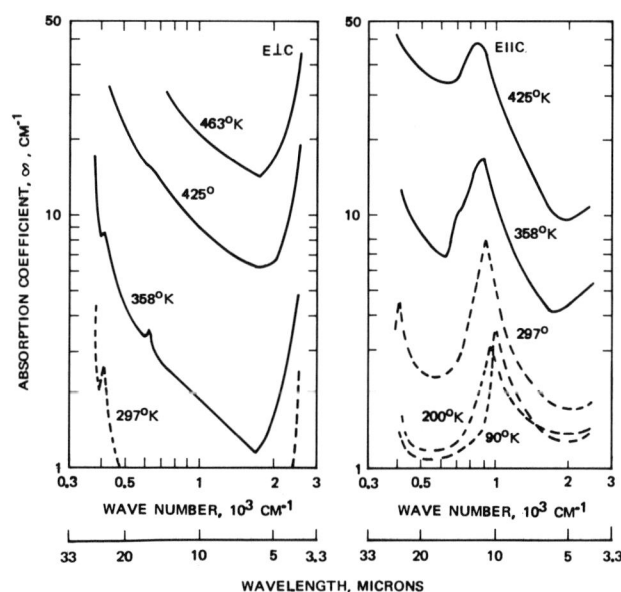

(REF. 4)

REFERENCES:

1. P. Billard, Acta Electronica, 6, 75-169, (1962).

2. P. Grosse, et al, Solid State Comm., 5, 99-100, (1967).

3. R.S. Caldwell and H.Y. Fan, Phys. Rev., 114, 664-675, (1959).

4. H.Y. Fan, Reports on Progress in Physics, 19, 107-155, (1956).

THALLIUM BROMOIODIDE

OPTICAL MATERIALS PROPERTIES DATA SHEET   MATERIAL: THALLIUM BROMOIODIDE (KRS-5)

INTRODUCTION: This data sheet describes the properties of mixed crystal thallium bromoiodide (KRS-5).

PHYSICAL PROPERTIES, (298°K)
- Density, (g/cm^3) 7.37
- Melting/Softening Temp. (°K) 688
- Solubility in Water, (g./100 g. H$_2$O) 0.05

THERMAL PROPERTIES, (298°K)
- Linear Expansion Coeff., (°K)$^{-1}$ 58×10^{-6}
- Thermal Conductivity (10^{-4} cal/(cm sec °K)) 13
- Specific Heat, (cal/g)/°K Not available

MECHANICAL PROPERTIES, (298°K)
- Young's Modulus, (psi) 2.3×10^6
- Hardness, (Knoop) 40.2

OPTICAL PROPERTIES, (298°K)

Dispersion Equation

$$n^2 - 1 = \sum_i \frac{K_j \lambda^2}{\lambda^2 - \lambda_j^2}$$

λ_j^2		K_j	
λ_1^2	0.0225	K_1	1.8293958
λ_2^2	0.0625	K_2	1.6675593
λ_3^2	0.1225	K_3	1.1210424
λ_4^2	0.2025	K_4	0.04513366
λ_5^2	27089.737	K_5	12.380234

Transmission Region, (External Transmittance ≥10% with 2.0 mm. thickness) 0.6 - 40µ

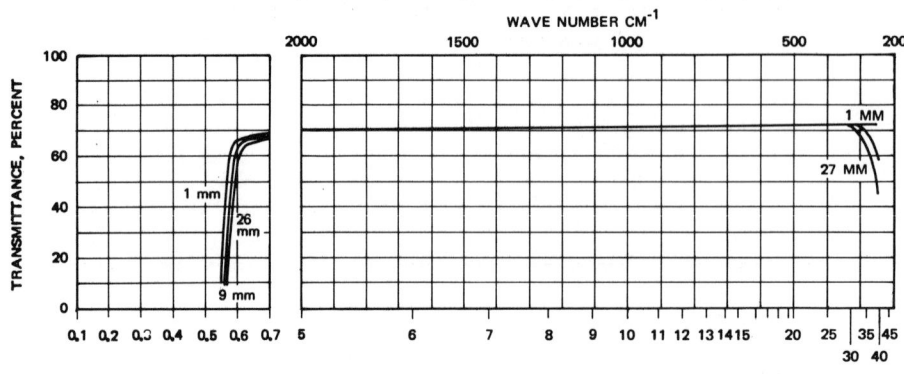

(REF. 1)

MATERIAL: THALLIUM BROMOIODIDE (KRS-5)

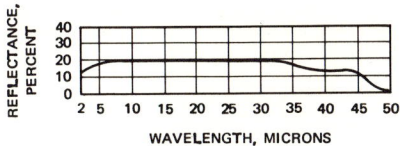
(REF. 2)

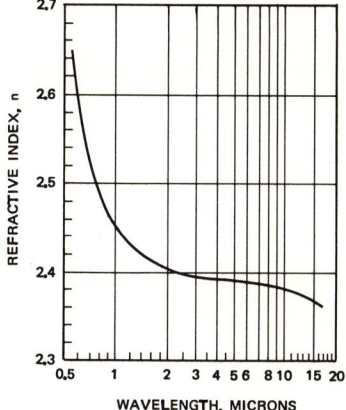
(REF. 3)

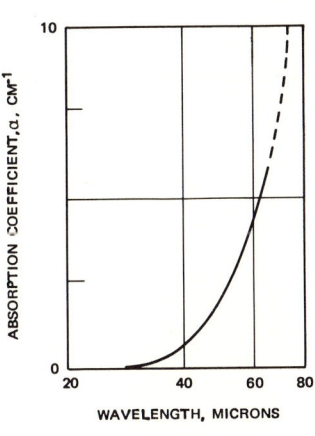
(REF. 4)

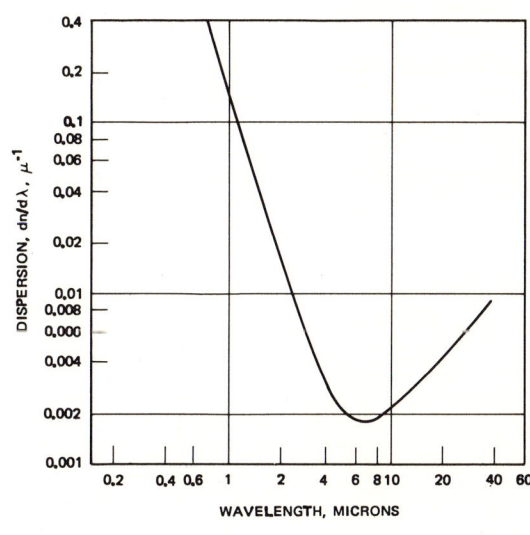
(REF. 5)

REFERENCES:

1. A. Smakula, et al, "Harshaw Optical Crystals," Harshaw Chemical Co., Cleveland, (1967).

2. D.E. McCarthy, Appl. Optics, 2, 591-595, (1963).

3. G. Joos, "F.I.A.T. Review of German Science, 1934-1946, The Physics of Solids," Part II, (1948).

4. A. Smakula, "Basic Properties of Optical Crystals with Special Reference to Infrared," Report No. PB111053, (1954).

5. A. Smakula, Opt. Acta., 9, 205-222, (1962).

THALLIUM CHLOROBROMIDE

OPTICAL MATERIALS PROPERTIES DATA SHEET

MATERIAL: THALLIUM CHLOROBROMIDE (KRS-6)

INTRODUCTION: This data sheet presents data for the mixed thallium chloride-thallium bromide crystal.

PHYSICAL PROPERTIES, (298°K)

Density, (g/cm^3)　　7.192

Melting/Softening Temp. (°K)　　697

Solubility in Water, (g./100 g. H$_2$O)　　0.32 (293°K)

MECHANICAL PROPERTIES, (298°K)

Young's Modulus, (psi)　　3.0×10^6

Hardness, (Knoop)　　~30 (500 g)

THERMAL PROPERTIES, (298°K)

Linear Expansion (Coeff., (°K)$^{-1}$　　50×10^{-6}

Thermal Conductivity (10^{-4} cal/(cm sec °K)　　17.1 (329°K)

Specific Heat, (cal/g)/°K　　0.0482

OPTICAL PROPERTIES, (298°K)

Dispersion Equation　　Not available

Transmission Region, (External Transmittance ≥10% with 2.0 mm. thickness)　　0.21 - 35µ

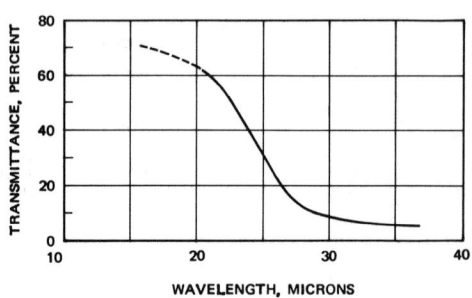

(REF. 1)

MATERIAL: THALLIUM CHLOROBROMIDE (KRS-6)

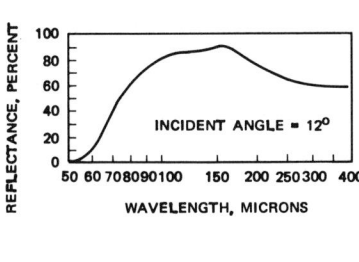

(REF. 2)

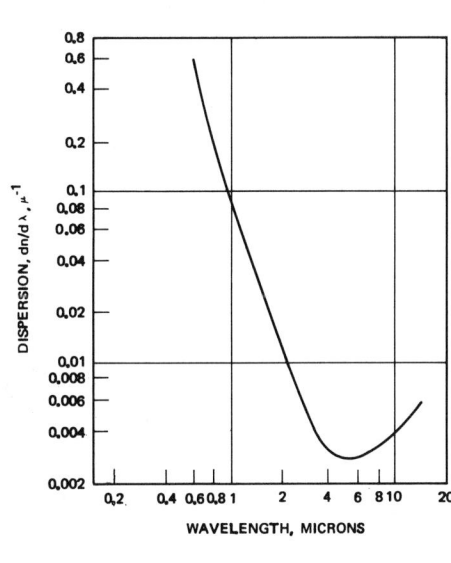

(REF. 3)

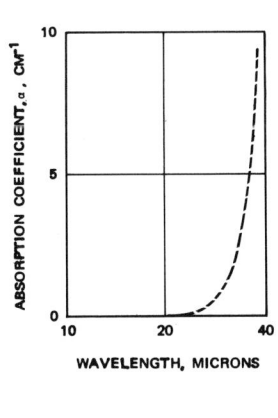

(REF. 4)

(REF. 4)

REFERENCES:

1. "American Institute of Physics Handbook," McGraw-Hill Book Co., New York, Second Ed., (1963).

2. A. Hadni, "Essentials of Modern Physics Applied to the Study of the Infrared," Pergamon Press, Oxford, (1967).

3. G. Hettner and G. Leisegang, Optik, 3, 305-314, (1948).

4. A. Smakula, Opt. Acta, 9, 205-222, (1962).

TITANIUM

OPTICAL MATERIALS PROPERTIES MATERIAL: TITANIUM
DATA SHEET

INTRODUCTION: This data sheet contains information on bulk titanium, except for film transmittance and reflectance data.

PHYSICAL PROPERTIES, (298°K)

Density, (g/cm^3)   4.50

Melting/Softening Temp. (°K)   2085

Solubility in Water, (g./100 g. H$_2$O)   <0.01

MECHANICAL PROPERTIES, (298°K)

Young's Modulus, (psi) 15×10^6

Hardness, (Brinell)   55 - 95

THERMAL PROPERTIES, (298°K)

Linear Expansion Coeff., (°K)$^{-1}$ 8.4×10^{-6}

Thermal Conductivity (10^{-4} cal/(cm sec °K)   490

Specific Heat, (cal/g)/°K   0.126

OPTICAL PROPERTIES, (298°K)

Dispersion Equation   Not available

Transmission Region, (External Transmittance ≥10% with _____ mm. thickness)   Not available

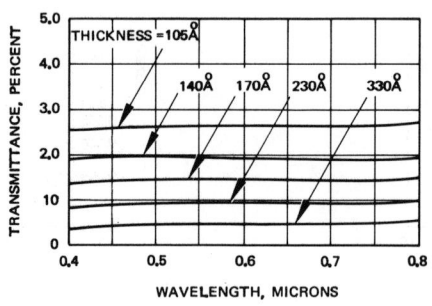

(REF. 1)

MATERIAL: TITANIUM

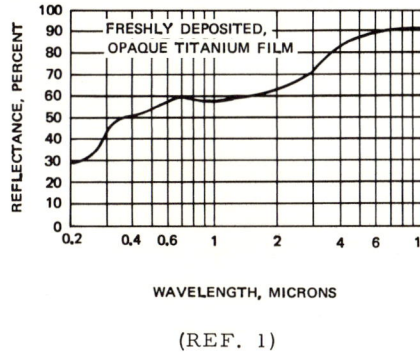

(REF. 1)

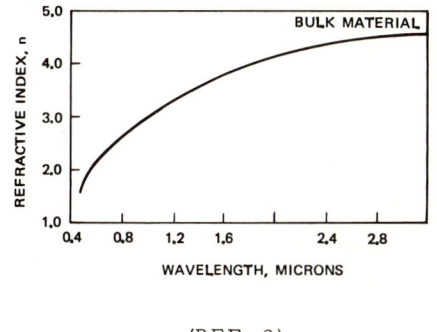

(REF. 2)

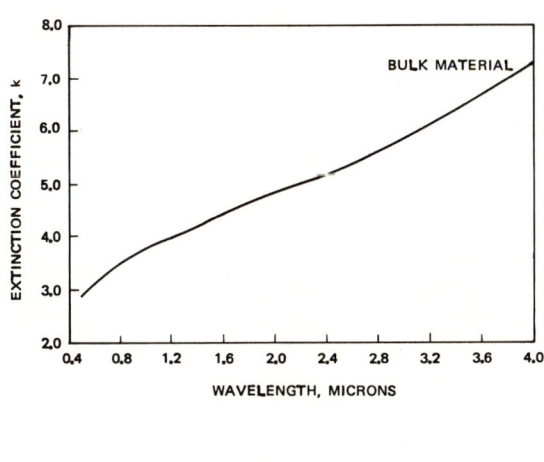

(REF. 2)

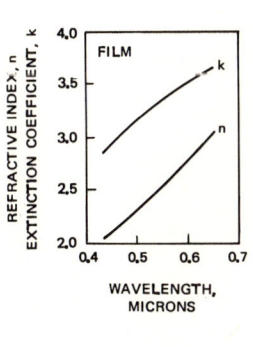

(REF. 1)

REFERENCES:

1. G. Hass and A.P. Bradford, J. Opt. Soc. Am., 47, 125-129, (1957).

2. M.M. Kirillova and B.A. Charikov, Phys. of Metals and Metall., 15, 138-139, (1963).

TITANIUM DIOXIDE

OPTICAL MATERIALS PROPERTIES
DATA SHEET

MATERIAL: <u>TITANIUM DIOXIDE (Rutile)</u>

INTRODUCTION: <u>This data sheet contains information for single crystal stoichiometric titanium dioxide.</u>

PHYSICAL PROPERTIES, (298°K)

Density, (g/cm^3)   4.26

Melting/Softening Temp. (°K)   2093

Solubility in Water, (g./100 g. H$_2$O)   <0.001

MECHANICAL PROPERTIES, (298°K)

Young's Modulus, (psi)   Not available

Hardness, (Knoop)   879 (500 g)

THERMAL PROPERTIES*, (298°K)

Linear Expansion Coeff., (°K)   (9.19 - 7.14) x 10^{-6}

Thermal Conductivity (10^{-4} cal/(cm sec °K))   300 ∥ C at 309°K, 210 ⊥ C at 317°K

Specific Heat, (cal/g)/°K   0.17

*Double values for E ∥ C and E ⊥ C respectively.

OPTICAL PROPERTIES, (298°K)

Dispersion Equation

Ordinary ray:

$$n^2 = 5.913 + 2.441 \times 10^7/(\lambda^2 - 0.803 \times 10^7)$$

Extraordinary ray:

$$n^2 = 7.197 + 3.322 \times 10^7/(\lambda^2 - 0.843 \times 10^7)$$

where λ = Å

Transmission Region, (External Transmittance ≥10% with <u>2.0</u> mm. thickness)   <u>0.43 - 6.2µ</u>

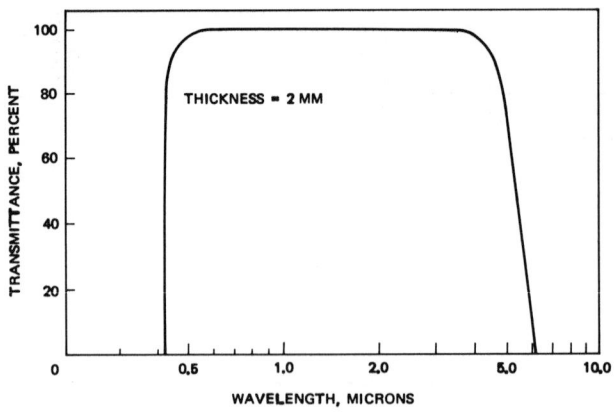

(REF. 1)

MATERIAL: TITANIUM DIOXIDE
(Rutile)

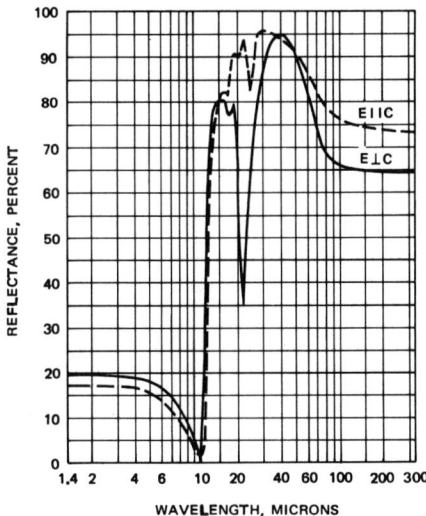

(REF. 2)

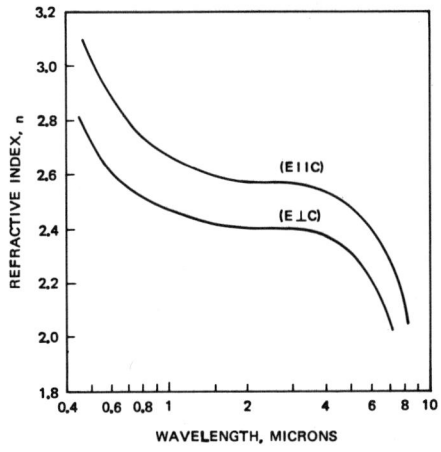

(REF. 3)

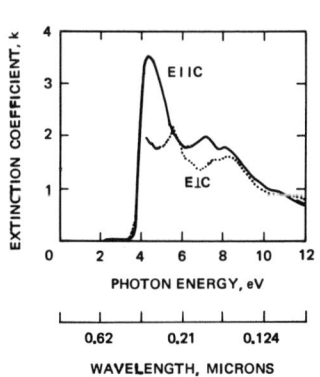

(REF. 4)

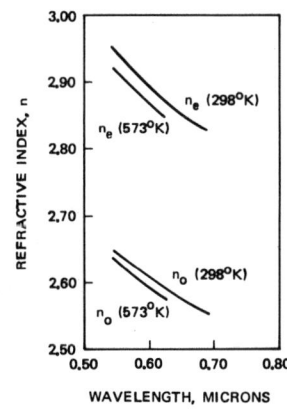

(REF. 5)

REFERENCES:

1. M.D. Beals and L. Merker, Materials in Design Eng., 51, No. 1, 12-13, (1960).

2. D.C. Cronemeyer, Phys. Rev., 87, 876-886, (1952).

3. D.C. Cronemeyer and M.A. Gilleo, Phys. Rev., 82, 975-976, (1951).

4. M. Cardona and G. Harbeke, Phys. Rev. 137, A1467-A1476, (1965).

5. A. Schroeder, Z. Kristallog, 67, 485-542, (1928).

ZINC SELENIDE

OPTICAL MATERIALS PROPERTIES DATA SHEET   MATERIAL: ZINC SELENIDE (Cubic)

INTRODUCTION: This data sheet contains data for cubic zinc selenide.

PHYSICAL PROPERTIES, (298°K)
- Density, (g/cm^3)   5.651
- Melting/Softening Temp. (°K)   1373
- Solubility in Water, (g./100 g. H$_2$O)   <0.001

MECHANICAL PROPERTIES, (298°K)
- Young's Modulus, (psi)   Not available
- Hardness, (Mohs)   3 - 4

THERMAL PROPERTIES, (298°K)
- Linear Expansion Coeff., (°K)$^{-1}$ 7×10^{-6}
- Thermal Conductivity (10^{-4} cal/(cm sec °K))   290
- Specific Heat, (cal/g)/°K   0.016

OPTICAL PROPERTIES, (298°K)
- Dispersion Equation   Not available
- Transmission Region, (External Transmittance ≥10% with 2.0 mm. thickness)   ~0.5 - $\overline{22}$μ

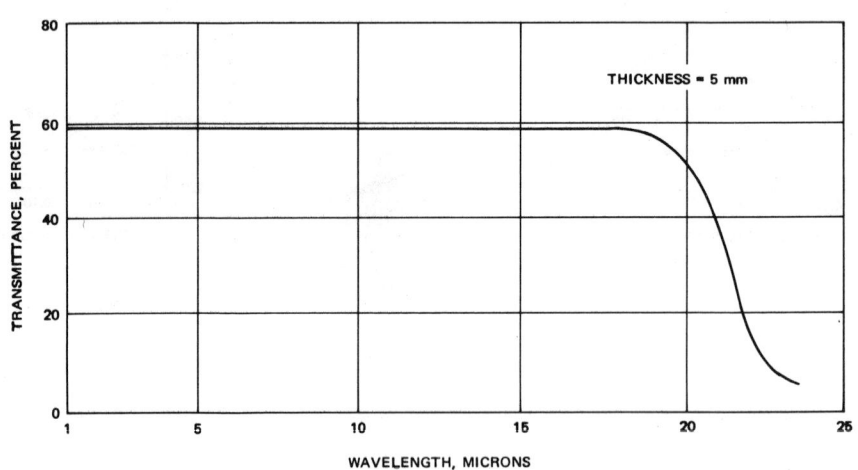

(REF. 1)

MATERIAL: ZINC SELENIDE
(Cubic)

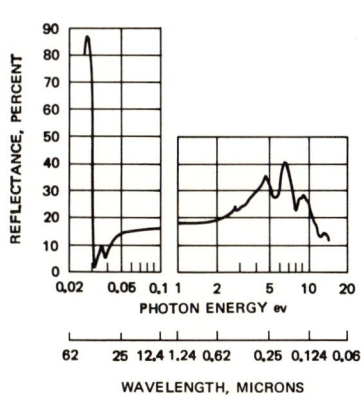

(REF. 2)

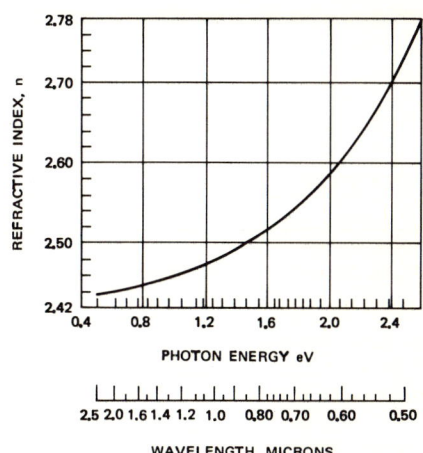

(REF. 3)

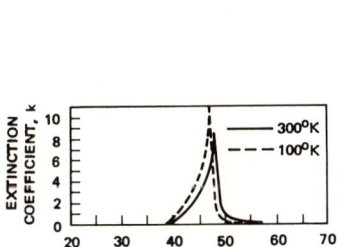

(REF. 4)

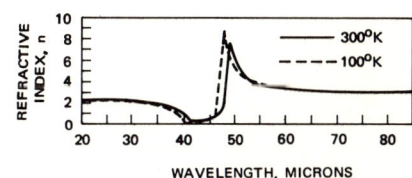

(REF. 4)

REFERENCES:

1. K.K. Dubenskiy, Soviet J. of Opt. Tech., 36, 118-121, (1969).

2. M. Aven, et al, J. Appl. Phys. Suppl., 32, 2261-2265, (1961).

3. D.T.F. Marple, J. Appl. Phys., 35, 539-542, (1964).

4. A. Manabe, et al, Japan, J. Appl. Phys., 6, 593-600, (1967).

ZINC SULFIDE

OPTICAL MATERIALS PROPERTIES MATERIAL: ZINC SULFIDE
DATA SHEET (Cubic)

INTRODUCTION: This data sheet presents information for single crystal cubic zinc sulfide.

PHYSICAL PROPERTIES, (298°K)
- Density, (g/cm^3) 4.09
- Melting/Softening Temp. (°K) 1293
- Solubility in Water, (g./100 g. H$_2$O) 6.5 x 10^{-5} (299°K)

MECHANICAL PROPERTIES, (298°K)
- Young's Modulus, (psi) 1.65 x 10^6
- Hardness, (Mohs) 3.5 - 4

THERMAL PROPERTIES, (298°K)
- Linear Expansion Coeff., (°K)$^{-1}$ 6.14 x 10^{-6} (273°K)
- Thermal Conductivity (10^{-4} cal/(cm sec °K)) 635
- Specific Heat, (cal/g)/°K 0.116

OPTICAL PROPERTIES, (298°K)

Dispersion Equation

$$n^2 = 5.164 + 1.208 \times 10^7/(\lambda^2 - 0.732 \times 10^7)$$

where λ = Å

Transmission Region, (External Transmittance ≥10% with 0.62 mm. thickness) ~0.6 - 15.6µ

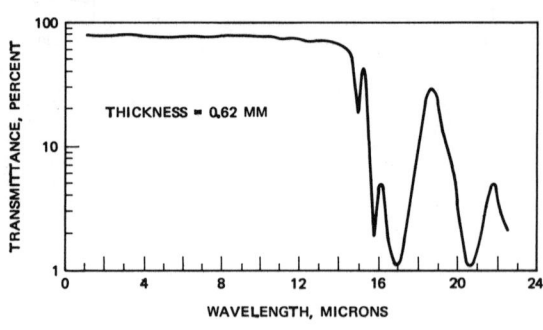

(REF. 1)

MATERIAL: ZINC SULFIDE
(Cubic)

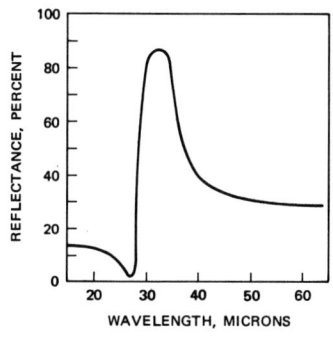

(REF. 2)

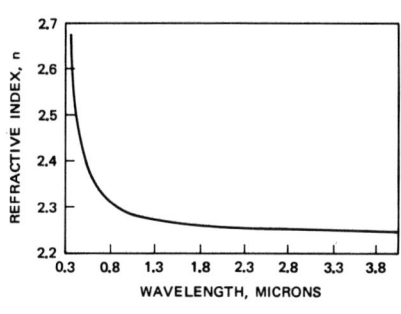

(REF. 3)

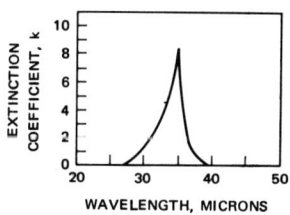

(REF. 2)

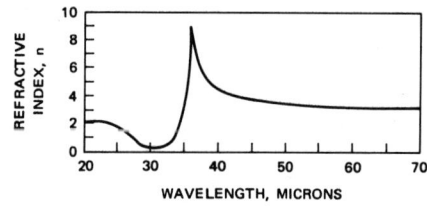

(REF. 2)

REFERENCES:

1. T. Deutsch, "Int. Conf. on the Phys. of Semiconductors, Proc. Exeter, 1962," A.C. Stickland, Ed., Inst. of Phys. & Phys. Soc., London, 505-512, (1962).

2. A. Manabe, et al, Japan. J. Appl. Phys., 6, 593-600, (1967).

3. S.J. Czyzak, et al, U.S. Government Report No. AD-143919, (1957).

APPENDIX A
WAVELENGTH CONVERSION FACTORS

(1) CONVERSION FACTORS FROM ELECTRONS VOLTS, (eV) to MICRONS (μ)

eV	μ	eV	μ	eV	μ
0.10	12.39	1.17	1.06	4.4	0.28
0.12	10.33	1.20	1.03	4.5	0.28
0.14	8.86	1.25	0.99	4.6	0.27
0.16	7.75	1.35	0.92	4.7	0.26
0.18	6.89	1.40	0.89	4.8	0.26
0.20	6.20	1.45	0.86	4.9	0.25
0.22	5.64	1.50	0.83	5.0	0.25
0.24	5.17	1.55	0.80	5.1	0.24
0.26	4.77	1.60	0.78	5.2	0.24
0.28	4.43	1.65	0.75	5.3	0.23
0.30	4.13	1.70	0.73	5.4	0.23
0.32	3.87	1.75	0.71	5.5	0.23
0.34	3.65	1.80	0.69	5.6	0.22
0.36	3.44	1.85	0.67	5.7	0.22
0.38	3.26	1.90	0.65	5.8	0.21
0.40	3.10	1.95	0.64	5.9	0.21
0.42	2.95	2.0	0.62	6.0	0.21
0.44	2.82	2.1	0.59	7.0	0.18
0.46	2.70	2.2	0.56	8.0	0.16
0.48	2.59	2.3	0.54	9.0	0.14
0.50	2.48	2.4	0.52	10.0	0.12
0.53	2.34	2.5	0.50		
0.56	2.21	2.6	0.48		
0.59	2.10	2.7	0.46		
0.62	2.00	2.8	0.44		
0.65	1.91	2.9	0.43		
0.68	1.82	3.0	0.41		
0.71	1.75	3.1	0.40		
0.74	1.68	3.2	0.39		
0.77	1.61	3.3	0.38		
0.80	1.55	3.4	0.36		
0.83	1.49	3.5	0.35		
0.86	1.44	3.6	0.34		
0.89	1.39	3.7	0.34		
0.92	1.35	3.8	0.33		
0.95	1.30	3.9	0.32		
0.98	1.26	4.0	0.31		
1.11	1.12	4.1	0.30		
1.14	1.09	4.2	0.30		
		4.3	0.29		

(2) CONVERSION FACTORS FROM WAVENUMBER, (cm^{-1}) to MICRONS, (μ)

Wavenumber cm^{-1}	μ	Wavenumber cm^{-1}	μ	Wavenumber cm^{-1}	μ
100	100	800	12.5	6500	1.54
110	90.9	825	12.12	7000	1.43
120	83.3	850	11.76	7500	1.33
130	76.9	900	11.11	8000	1.25
140	71.4	925	10.81	8500	1.18
150	66.7	950	10.53	9000	1.11
160	62.5	975	10.26	9500	1.05
170	58.8	1000	10.00	1×10^4	1.00
180	55.6	1050	9.95	1.1	0.91
190	52.6	1100	9.09	1.2	0.83
200	50.0	1150	8.70	1.3	0.77
220	45.5	1200	8.33	1.4	0.72
240	41.7	1250	8.00	1.5	0.67
260	38.5	1300	7.69	1.6	0.62
280	35.7	1350	7.45	1.7	0.59
300	33.3	1400	7.14	1.8	0.56
320	31.2	1450	6.90	1.9	0.53
340	29.4	1500	6.67	2.00	0.50
360	27.8	1550	6.45	2.1	0.48
380	26.3	1600	6.25	2.2	0.46
400	25.0	1650	6.06	2.3	0.42
420	23.8	1700	5.88	2.4	0.38
440	22.7	1750	5.71	2.5	0.37
460	21.7	1800	5.56	3.0	0.33
480	20.8	1850	5.41	3.5	0.28
500	20.0	1900	5.26	4.0	0.25
520	19.2	1950	5.13	4.5	0.22
540	18.5	2000	5.00	5.0	0.20
560	17.9	2200	4.55	6.0	0.17
580	17.2	2400	4.17	7.0	0.14
600	16.7	2600	3.85	8.0	0.12
620	16.1	2800	3.57	9.0	0.11
640	15.6	3000	3.33	1×10^5	0.10
660	15.2	3500	2.85		
680	14.7	4000	2.50		
700	14.3	4500	2.22		
725	13.8	5000	2.00		
750	13.3	5500	1.82		
775	12.9	6000	1.67		

APPENDIX B
GLOSSARY OF OPTICAL TERMS

ABSORPTION COEFFICIENT, (α) - defined by the equation:

$$I_x = I_o e^{-\alpha x}$$

where I_o = incident radiation intensity

I_x = transmitted radiation intensity at distance of x cm into the material.

α = absorption coefficient, cm^{-1}

BIREFRINGENCE (Double Refraction) - division of incident light by optically anisotropic materials into two component which are diffracted in different directions.

DISPERSION, ($dn/d\lambda$) - the derivative of the refractive index with respect to wavelength,

where: λ = wavelength, microns.

EXTINCTION COEFFICIENT, (k) - defined by the equation:

$$k = \frac{\alpha \lambda}{4\pi}$$

where λ = wavelength in cm
α = absorption coefficient
k = extinction coefficient, dimensionless

EXTRAORDINARY RAY - wave traveling with a velocity which depends on the relation between its direction and the optic axis, being polarized with its electric vector parallel ($\parallel$) to the plane containing the optic axis and the direction of propagation. (See also birefringence)

ORDINARY RAY - wave traveling with a velocity independent of the direction of propagation, being polarized with its electric vector perpendicular ($\perp$) to the direction of the optic axis and the direction of propagation. (See also birefringence)

REFLECTANCE, (R) - The percentage of the incident light reflected by a surface.

REFRACTIVE INDEX, (n) - the ratio of the phase velocity of light in vacuum to the phase velocity in the medium.

TRANSMITTANCE, (T) - the percentage of incident light transmitted through the material and observed. (also called external transmittance)

QC
176.8
O 6
M 67

SEP 11 1972